What Your Colleagues Are Saying...

This book is for any educator, parent, and activist who understands that education is not neutral, for any citizen of this world who believes that a quality education allows its recipients to understand, thrive, and transform their world. An education thusly positioned changes lives and benefits humanity.

Lisa Williams
Partner, Equity in Education
Baltimore, MD

Teaching Mathematics for Social Justice, Grades K-12: A Guide for Moving From Mindset to Action is a thoughtfully designed guide that helps math educators know their children, use practices to highlight their strengths, and ensure the math they teach is relevant. This book will assist you in designing experiences for children to learn mathematics skills and concepts and use them to make the world a better and more just place! Begin your journey as a culturally relevant educator by taking this guide and using it to reimagine your math classroom.

Georgina Rivera
NCSM Vice President and Principal
West Hartford, CT

In a climate where equity and diversity efforts are under attack, Childs and Staley courageously advocate for teaching mathematics for social justice. Their framework centers children, incorporates standards-based mathematics instruction as well as culturally relevant and responsive teaching, and has equity as its foundation. This book is a must read for K-12 teachers who want to engage in continuous improvement of their instructional practices.

Kyndall Brown
Executive Director, California Mathematics Project
Los Angeles, CA

Teaching Mathematics for Social Justice, Grades K-12: A Guide for Moving From Mindset to Action is a book that helps educators develop a framework for teaching mathematics with a focus on social justice. Each chapter of the book includes a "Reflect and Action" box, which guides readers to think critically throughout their journey toward teaching mathematics for social justice. Overall, this book is a useful resource for those who are committed to promoting social justice in their mathematics teaching practices.

Robert Q. Berry III
Dean & Professor, College of Education, University of Arizona
Tucson, AZ

Enhance your justice journey with this excellent resource. Offering accessible and actionable guidance to effectively integrate a Teaching Math for Social

Justice approach in K–12 classrooms, this book includes practical tools and tasks to plan, enact, and reflect on instruction that centers children, cultivates joy, and sparks action to make a positive difference in our world with mathematics.

Julia Aguirre
Professor, University of Washington Tacoma
Renton, WA

Childs and Staley have created a practical and powerful resource for teachers who are looking to design inclusive classrooms and working to ensure that ALL students have access to rigorous and important mathematics instruction. This will be a valuable resource that I will undoubtedly return to for guidance and support as I continue to engage in this work.

Zak Champagne
Lead Teacher, The Discovery School
Jacksonville, FL

Teaching Mathematics for Social Justice, Grades 6–12: A Guide for Moving From Mindset to Action offers a transformative framework for educators to cultivate equity in the classroom. From understanding personal biases to engaging communities, this guide provides practical strategies to empower teachers and students alike, making it a must-read for anyone committed to social justice in education.

Pamela Seda
Co-author, *Choosing to See: A Framework for Equity in the Math Classrooms*
Atlanta, GA

Childs and Staley have developed an engaging and easy-to-read practitioner's blueprint for teaching mathematics for social justice by providing entry points for everyone, no matter the level in which they already engage. Thoughtfully planned examples, reflections, and next steps create for a practical learning experience for educators.

Crystal M. Watson
Consultant, Crystal M. Watson Consulting
Cincinnati, OH

Teaching Mathematics for Social Justice

Grades K–12

Teaching Mathematics for Social Justice

Grades K–12
A Guide for Moving From Mindset to Action

Dr. Kristopher J. Childs
Dr. John W. Staley

CORWIN **Mathematics**

For information:

Corwin
A SAGE Company
2455 Teller Road
Thousand Oaks, California 91320
(800) 233–9936
www.corwin.com

SAGE Publications Ltd.
1 Oliver's Yard
55 City Road
London, EC1Y 1SP
United Kingdom

SAGE Publications India Pvt. Ltd.
Unit No 323–333, Third Floor,
F-Block
International Trade Tower Nehru
Place
New Delhi – 110 019
India

SAGE Publications
Asia-Pacific Pte. Ltd.
18 Cross Street #10–10/11/12
China Square Central
Singapore 048423

Vice President and Editorial Director:
    Monica Eckman
Associate Director and Publisher,
    STEM: Erin Null
Senior Editorial Assistant:
    Nyle De Leon
Production Editor: Tori Mirsadjadi
Copy Editor: Diana Breti
Typesetter: Integra
Proofreader: Lawrence W. Baker
Indexer: Integra
Cover Designer: Scott Van Atta
Marketing Manager:
    Margaret O'Connor

Printed in the United States of America.

Paperback ISBN 978-1-0718-4694-0

This book is printed on acid-free paper.

24 25 26 27 28 10 9 8 7 6 5 4 3 2 1

CONTENTS

Visit the companion website at
https://qrs.ly/wbfixtr
for downloadable resources.

Note From the Publisher: The authors have provided video and web content throughout the book that is available to you through QR (quick response) codes. To read a QR code, you must have a smartphone or tablet with a camera. We recommend that you download a QR code reader app that is made specifically for your phone or tablet brand.

ACKNOWLEDGMENTS

Dr. Childs's Acknowledgments

I would like to take a moment to acknowledge and thank four key people who played a major role in my life and made me into the Black man I am today. During the writing of this book between 2020 and 2021, all four became ancestors. Their individual spirits gave me what I needed to complete this book. Thus, I would like to honor and acknowledge my grandfather Clarence Zinnerman, my grandmother Patsy Harris, my close friend and fraternity brother Dr. Damien Moses, and my father Obie J. Childs. I love you all and thank you for everything!

Dr. Staley's Acknowledgments

I want to thank my wife, Karen, for walking beside me and continuing to encourage me to follow God's plan. To our children, Jonathan, Alexis, and Mariah, keep chasing your dreams and changing the world. Bevelyn and Asa, you are the reason I continue this journey.

To the many mathematics leaders, mathematics teachers, educators, colleagues, and friends, thank you for the opportunity to work with and walk with you to make mathematics more meaningful, relevant, and accessible for each and every student. *Together we are changing the narrative!*

Publisher's Acknowledgments

Corwin gratefully acknowledges the contributions of the following reviewers:

Kyndall Brown
Executive Director, California Mathematics Project
Los Angeles, CA

Zak Champagne
Lead Teacher and Director of Mathematics and Diversity, Equity, and Inclusion,
The Discovery School
Jacksonville, FL

Tashana Howse
Professor of Mathematics Education, Georgia Gwinnett College
Lawrenceville, GA

Ishmael Robinson
Math Instructor, University of Minnesota
Grove Heights, MN

Crystal M. Watson
Consultant, Crystal M. Watson Consulting
Cincinnati, OH

ABOUT THE AUTHORS

As a teacher, professor, consultant, and member of senior leadership teams, Dr. **Kristopher J. Childs** focuses on excellence in teacher content and pedagogical knowledge, equity, leadership development, and organizational change. His work is guided by his mantra, "Live life to the fullest; you only get one."

Dr. Childs is a highly sought-after keynote speaker known for his coaching skills, storytelling, and passion. As a speaker, Dr. Childs inspires audiences to relentlessly pursue their goals and dreams. He seeks to help each find his or her passion and purpose, and his messages have been deemed "life-changing" by clients.

Dr. Childs seeks to create a movement through educating, advocating, and inspiring individuals to pursue academic excellence. Due to his student-centered approach to teamwork, faculty, staff, and students recognize him as a visionary and collaborative leader. Dr. Child's approach helps both school and district teams that he consults achieve common goals and improve student academic success and students' classroom experiences.

Dr. Childs is a life member in the Florida Agricultural and Mechanical University National Alumni Association and a life member in the Omega Psi Phi Fraternity, Incorporated. He earned his doctorate in mathematics education from the University of Central Florida, his master of science degree in mathematics education from Nova Southeastern University, and his bachelor of science degree in computer engineering from Florida Agricultural and Mechanical University.

He is a co-author of *Making Sense of Mathematics for Teaching Girls in Grades K-5* and author of the article "Good Mathematics Teaching is NOT Telling, It Is Facilitating." Follow him on all social media platforms—@DrKChilds and www.kristopherchilds.com—to learn more about his work.

John W. Staley, PhD, has been involved in mathematics education for more than 35 years as a secondary mathematics teacher, adjunct professor, district and national leader, and consultant. During his career he has presented at state, national, and international conferences; served on many committees and task forces; facilitated workshops and professional development sessions on a variety of topics; and received the Presidential Award for Excellence in Teaching Mathematics and Science. A past president for NCSM, the mathematics education leadership organization, and past chair of the U.S. National Commission on Mathematics Instruction, he continues to serve on several advisory boards and is a co-founder of Math Milestones. He is a co-author for *Middle School* (2023) and *High School* (2022) *Mathematics Lessons to Explore, Understand, and Respond to Social Injustice* (Corwin/NCTM); *Catalyzing Change in High School Mathematics: Initiating Critical Conversations* (NCTM); and *Framework for Leadership in Mathematics Education* (NCSM). John's current passion and work focus on changing the narrative about who is seen as being doers, learners, and teachers of mathematics, especially for African American boys and men; student readiness for algebra and success during the transition years; and building mathematics education leaders at all levels. Follow him on X—@jstaley06—to learn more about his work.

INTRODUCTION

Think back to the first time you taught a math lesson. For many of us, that day was filled with anticipation, excitement, nervousness, and we're sure a variety of other emotions as you greeted that first group of eager young math minds. Before you knew it, day one was done. You may still have many mixed emotions as you remember everything: the grade level or courses you were assigned to teach, the colleagues who worked beside you, the school and community you were embedded in, and the children you taught. Whether you knew it at the time or not, this was the beginning of your social justice journey. For many teachers, this journey began the first time they became aware that some children were not achieving at similar levels as their peers; some children were not included in or seemed to be missing from certain mathematics classes. Maybe they noticed discrepancies when they witnessed some classrooms in which students experienced rich, meaningful, engaging, collaborative learning and contrasted this with other classrooms in which students were expected to complete volumes of worksheets or textbook pages focused on repeatedly solving examples of mathematical procedures, with maybe a few word problems sprinkled in. Maybe they were involved in or overheard conversations about "what those children could not do" or "how those families did not show up."

At some point, they became aware that student achievement data could be predicted based on student characteristics and which "class" students were in. They may have acknowledged that the pattern of low performance was persistent for some children and they had a growing feeling that there had to be more they could do for the children in their classes, school, and community so that these children would have a more meaningful, relevant, and fulfilling experience in the mathematics classroom.

Pause for a moment and really think back. Was this you? Was this the start of your own Teaching Mathematics for Social Justice journey?

If you have picked this book up, it is because you

> see injustices in your educational setting and/or your community;

> see how children's identities impact their educational experience;

> view students as youthful and innocent children to whom you owe the best mathematical learning experience you can offer;

> believe every child should receive a high-quality educational experience;

> are a committed educator who wants to focus on teaching mathematics in a way that is equitable and provides access to *all* children; and

> are an unapologetic social justice educator focused on disrupting, dismantling, and re-imagining children's mathematical education experiences. No exceptions.

If you see, feel, or are any of these things, it is likely you are at least curious, if not committed, to looking for ways to (1) recognize and name the injustices you see in your school or district, (2) modify or expand your teaching practice in order to reach more children and provide more equitable learning experiences for all of them, and (3) disrupt and dismantle systems of oppression and inequity you see play out in K–12 education. In other words, you are someone who is on a journey to becoming a Social Justice Mathematics Educator. If this describes you, you are in the right place. You are embarking on a meaningful and life-changing journey that can help you reinvigorate your professional purpose and may even have an influence on your personal life.

You are embarking on a meaningful and life-changing journey that can help you reinvigorate your professional purpose and may even have an influence on your personal life.

What This Book Is About

This book will lead you through a transformational journey to reimagine your mathematics classroom in a way that not only better serves more children but also lifts mathematics up as a tool for them to analyze and understand the worlds around them, celebrate their unique identities and their communities, and become agents of change. It introduces the inherent inequities in children's educational experiences and will help you learn how to address these inequities by providing children with meaningful, relevant, and liberatory mathematics learning experiences that connect to social issues that affect children's lives and extend beyond the classroom and into their community. You will be challenged to reimagine the classroom mathematics experience and harness the potential children bring into the mathematics classroom. We call this teaching mathematics for social justice (TMSJ).

As you proceed, your focus should be on the paths to be taken, not on the destination. Along the way, you will refine your understanding of mathematical concepts and pedagogy. You will also become better able to understand the academic and instructional needs of diverse groups of children and improve your understanding of the different communities from which they come. This book will challenge you to rethink your paradigms and beliefs about the practice and purpose of mathematics instruction: What is teaching mathematics for social justice? Who are you and how do you show up in your math class? Who are your children? And what are some of the best ways to teach mathematics? Most importantly, you will receive practical techniques and examples so you can implement TMSJ in your classroom.

During your journey to becoming a social justice mathematics educator, you may at times feel supported, challenged, alone, and validated. You will be asked to take a critical look in the mirror at yourself and your practices. It may feel uncomfortable and personal; thus, it is important that you take the time you need to self-reflect. This book is designed to challenge and push your thinking but also to give you the time and space needed to expand your critical consciousness of your mathematics instruction—considering who it serves and how it serves them—in a way that is safe. You may, at times, experience cognitive dissonance, and you may even sometimes feel defensive. But all good educators know that it is only through that dissonance and vulnerability that we grow. The key is to be gentle but honest with yourself and keep pushing forward. If you falter, make a mistake, or feel stuck at some point, that is OK. It is to be expected, but you can and will get up and keep growing.

It's also important to recognize that this journey never ends. If done well, your journey will last throughout your career. You can start with the small steps this book describes and encourages, and you'll gradually be able to take bigger and bigger leaps. Along the way, you'll be shaping the hearts and minds of the children in your care, helping to empower them to use their own mathematical brilliance to become agents of change. Just remember, the best part of any journey is the journey itself, not the destination. As you embark on the journey through the four parts in this book, you will find that the chapters in each offer practical stepping stones and actions you can take to help you better understand yourself and serve your children, make sense of their values and their communities, refine the concept of mathematics, rethink your paradigm and beliefs, critically evaluate your mathematical instructional practices and resources, plan and enact a social justice mathematics experience, and learn to collaborate with students, other education professionals, and outside collaborators as you build a community dedicated to helping children use mathematics to leave a lasting change.

Who This Book Is For

We wrote this book for K-12 teachers, coaches, administrators, and preservice teachers because each plays a critical role in children's mathematical learning and engagement:

▶ Teachers help authentically engage children and bring meaning to mathematics lessons. Teachers who want to move beyond compliance-based lessons that feel irrelevant to children and move toward lessons with a greater impact on meaningful learning will benefit from this book.

▶ Coaches help guide the creation of impactful mathematics experiences that prepare children for life outside of school and help them see how to use mathematics to understand the world.

▶ Administrators understand inequities occurring under their purview with a focus on improving student achievement and understanding that the opportunity gap as status quo is no longer acceptable.

▶ Preservice educators prepare their students not for the classroom that exists but the classroom that is on the horizon by helping develop a lens of understanding for a population of children destined for greatness.

How This Book Works

Because we think of becoming a social justice educator as a journey, this book is laid out accordingly. The journey consists of four main parts, each of which includes several chapters (see Figure i.1).

▶ *Part 1 A Social Justice Mathematics Teaching Framework* starts the journey by introducing TMSJ, as we pause to look in the mirror (self-reflect) and look through the windows at our children. Chapter 1 focuses on making sense of TMSJ. Chapter 2 then provides you with an opportunity to reflect on who you are as you enter this work, and in Chapter 3 we take time to learn about the children we serve and who will be impacted by this work.

▶ *Part 2 The Mathematics Experience* continues the journey with a focus on the ideal mathematical environment for TMSJ. Chapter 4 introduces the cornerstones for creating an ideal mathematics environment that centers your children's culture, fosters a sense of community, and encourages collaboration. Chapter 5 looks at the fourth cornerstone, the importance of engaging all children so that they can reach their potential.

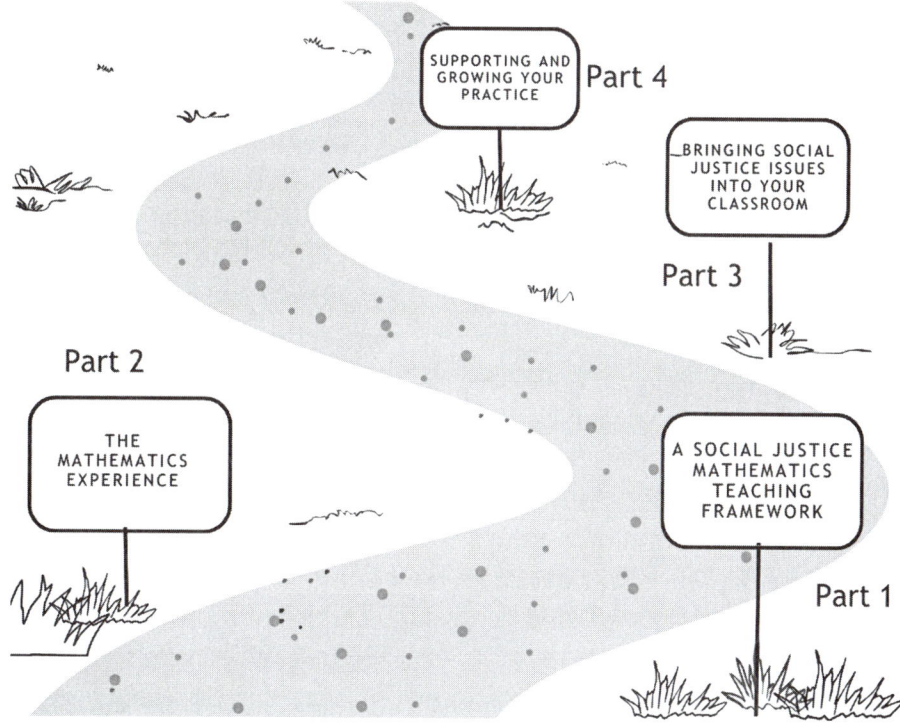

i.1 Your TMSJ Journey

Part 4 — SUPPORTING AND GROWING YOUR PRACTICE

Part 3 — BRINGING SOCIAL JUSTICE ISSUES INTO YOUR CLASSROOM

Part 2 — THE MATHEMATICS EXPERIENCE

Part 1 — A SOCIAL JUSTICE MATHEMATICS TEACHING FRAMEWORK

▶ *Part 3 Bringing Social Justice Issues Into Your Classroom* focuses on bringing TMSJ to life through making sense of social justice issues and incorporating them into the classroom. Chapter 6 focuses on the practical keys of how a TMSJ lesson comes together, from selecting a social justice issue to planning for supports for your children to access the mathematics and context of the social issue. We then walk you through the implementation of a social justice mathematics lesson with examples in Chapter 7 and wrap up taking TMSJ to the next level by considering how to take action beyond the classroom in Chapter 8.

▶ *Part 4 Supporting and Growing Your Practice* concludes the journey with building a community of collaborators and a deep reflection about where you can go from here. Chapter 9 explores the various roles collaborators—children, families, other educators, and community members—might play in your journey and the benefits of growing a community of collaborators to walk beside you. We wrap up the journey in Chapter 10 by looking back at your "Why" and looking forward to the future and the legacy you are charting on your TMSJ journey.

Throughout the journey, features are included to provide practical next steps to begin incorporating TMSJ in your classroom and time to reflect upon what you want for yourself and your children. Each feature provides opportunities for you to interact with the chapter and document your insights, wonderings, and noticings along the journey as you begin to make changes, either small scale or large scale, to move forward in the process of creating high-quality mathematics experiences for all the children in your classrooms and to engage in meaningful and sustainable action.

▶ **Check-In:** This feature provides you with an opportunity to engage in a quick practical activity to apply what you have learned in the chapter.

▶ **Try This:** This feature consists of activities to help you apply the chapter content to your local setting: self work, your classroom, your school, or your community.

▶ **Reflect and Act:** This feature is included at the end of each chapter so that you can reflect on the TMSJ techniques and strategies discussed. In **Reflect** a series of rhetorical questions are provided so you can recall key takeaways from the chapter and consider how the information shared in the chapter applies to you, your children, and your mathematics environment. **Act** calls on you to document those key takeaways and next steps for continued learning and application to extend your thinking and enhance your work. In addition, activities similar to Try This are included to help you apply what you have learned in the chapter and begin to make changes.

▶ **The TMSJ Action Plan:** This is a tool that provides space for you to record your key takeaways from each of the chapters and Next Steps to support you as you continue through each stage of your social justice mathematics teaching journey. This tool is available for download at https://qrs.ly/wbfixtr

▶ **Where to Next:** This section serves as a bridge between chapters and provides you a preview of the next chapter's content.

▶ **Conversations With Dr. Childs and Dr. Staley:** QR codes link to audio recordings of the authors sharing their thinking as they navigated the writing of this book. You will gain insights into their takeaways, challenges they are still thinking about, and hope for future work as they continue their TMSJ journey.

How You Can Use This Book

This book is not designed to be read cover to cover and then placed on a bookshelf as a reference book. *Teaching Mathematics for Social Justice, Grades K-12: A Guide for Moving From Mindset to Action* was intentionally designed for reflection, interactivity, and action. Use this as a guidebook to help unpack who you are as an educator while simultaneously learning the tools necessary to transform your mathematics environment. Teachers can use it individually, in a professional learning community, or within a book study. In addition, this book can be used with preservice teachers who are seeking to enhance their instructional strategies as they prepare for the classroom. This book is not designed to be a quick over-the-week-end read; however, it is designed to expand your thinking and help you evolve as an educator. Give yourself ample time to work through the book—including the reflection and application elements—and make sense of the material being presented. Finally, the key to having a successful experience with this book is keeping a child-centered mindset and continually reflecting on the question "Do children in schools deserve better mathematics education experiences?"

> The key to having a successful experience with this book is keeping a child-centered mindset and continually reflecting on the question "Do children in schools deserve better mathematics education experiences?"

A Note About Language

Throughout this book, we have been very deliberate about the language we choose to use. Language has power, is fluid, and is ever-changing. Language can be a vehicle for change; it can reinforce power dynamics and forms of domination, or it can interrupt them. This is especially true in the ways we talk about historically excluded identities pertaining to race, ethnicity, socioeconomic status, and gender. These identities are being intentionally highlighted as, across the United States and other countries, researchers continue to use them to explain the consistent achievement gaps (Carnoy & García, 2017).

Allow us a moment to explain our choices: First, we use the word *children* instead of *students* to describe the young people we encounter in our classrooms and schools. This is intentional, as it conveys our strong belief in the innocence and humanity of the young persons served by the preK-12 educational system, approximate ages 4–18, in North America and in other countries. When people use the term *students* in the formal sense to describe someone who attends school, it can create a dissonance between children's identities as learners and their identities as smaller, or at least younger, human beings. Also, in many schools, students are recognized by identification numbers only, a reality that further disassociates children from their personhood. So, let's be abundantly clear: Throughout this book, we refer to a person or persons who attend school as a *child* or as *children* for the following reasons:

▶ The terms *child* and *children* humanize the very real and unique persons we educators are charged to teach.

▶ Every child who attends school is (or was once) *someone's* pride and joy. Therefore, as teachers, we need to value them just as much as their communities, families, and other loved ones do.

▶ A person is always considered a child to his, her, or their family or community elders, whether he, she, or they are the youngest of school attendees or about to graduate from high school.

▶ Given their innate youthfulness, innocence, and humanity, *all* children are deserving of life, love, and respect—in and out of the school setting.

Further, we use the term *Black* as a more inclusive term and to recognize there are Black people all over the world. We use the term in the context of race and culture and to center Black people's place in the historical narrative of the universe.

Melanated and *non-melanated* will be used in the book to decenter white-ness further. Often folx use the phrase "people of color" to describe non-white-identifying people. However, this phrase centers whiteness. We must consider that globally, at least three-quarters of the people are non-white and "people of color" subjects the global majority to a white-centered lens (Welsing, 2004). Thus, the intentional use of the term *melanated* to refer to folx who have high concentrations of melanin in their skin properly acknowl-edges their place in a global view. In addition, *non-melanated* will be used to refer to folx who identify as white.

Folx will be used to intentionally and explicitly signal the inclusion of groups commonly or historically excluded (Merriam-Webster, n.d.). In the work of TMSJ we must have a keen focus on historically excluded groups of folx, which further signifies a decentering of whiteness within the work.

Environments will be used as an all-inclusive term when referring to anything related to a learning location; that is, it is not limited to the traditional "in the building" classroom. *Environments* can also refer to the school building itself, the community, the home—essentially anywhere a child can gain knowledge and information.

This book will also have a keen focus on strengths-based language, as opposed to deficit-based language. If we are going to reimagine mathematics educa-tional experiences, a key component is rethinking the language we use in edu-cation. Often, stakeholders use a deficit lens when children aren't from the majority group. For example, the term *minority* is used to describe non-white children. We often place children in student groups with deficit-based descriptors, such as English Language Learners (which centers English) and students with disabilities (as opposed to focusing on their abilities). Collectively, we as mathematics educators need to work to reframe our use of language. This book will be a tool in that process.

This book is an all-encompassing journey designed to challenge the status quo mindset of the current mathematics education experience. Keep an open mind as you embark upon this journey and allow yourself to truly experience a shift as you navigate each part and journal your insights. Some of the information presented will serve as confirmation; for others, the information presented will be brand new. However, all the information presented will be impactful and help further your journey of becoming a social justice–focused mathematics educator. This book may not be for everyone; it is intentionally designed for those who see what children currently experience in mathematics education and desire something better. You are that educator. Let's begin the journey by first making sense of the social justice mathematics teaching framework.

PART 1

A SOCIAL JUSTICE MATHEMATICS TEACHING FRAMEWORK

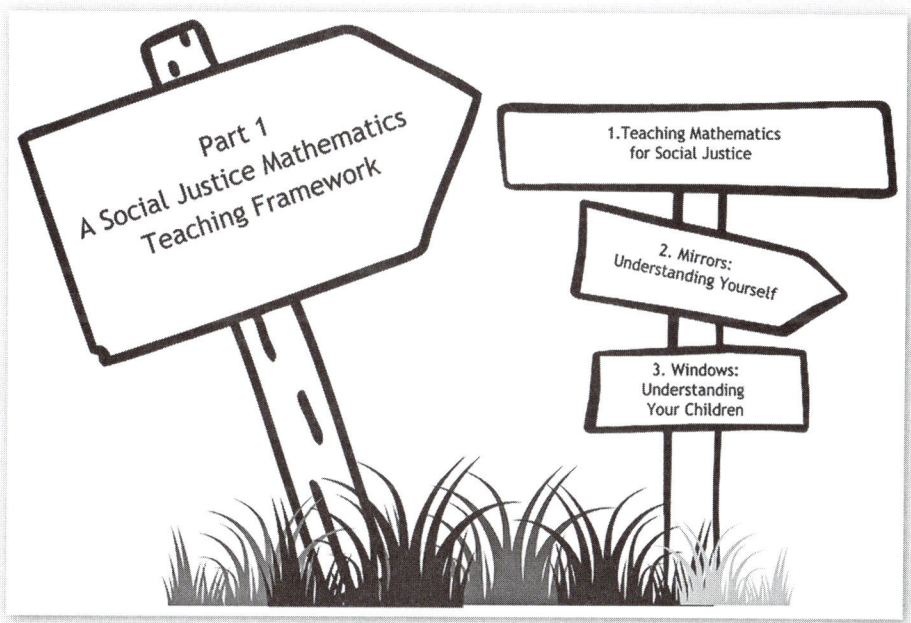

Your journey to becoming a social justice mathematics educator begins here. Before you can enact change in your educational environment, it's critical to first understand some of the foundations of what teaching mathematics for social justice (TMSJ) is and why it matters (Chapter 1). It is equally important to situate yourself in this work and take a deep look internally to learn what drives you in this work and in your worldview (Chapter 2), as well as what drives your children and how you can come to understand them better (Chapter 3). These are the goals of Chapters 1 through 3. Let's take the first step on the journey.

TEACHING MATHEMATICS FOR SOCIAL JUSTICE: WHAT IS IT AND WHY DOES IT MATTER?

Every journey has a starting point where you take a moment to set your sights on your next destination carefully. As mathematics educators navigating social justice, we realize that each of us has our own beginning point, reasons for taking this journey, and destination, so we hope that this guide will help you navigate to your "next destination." In this chapter, our journey begins as you

> imagine a vision of what equity looks like in the mathematics classroom,

> explore the social justice mathematics teaching framework, and

> consider the components of becoming a social justice mathematics educator.

A Vision of Equity in the Mathematics Classroom

As 21st century mathematics educators, we must begin reimagining what mathematics looks and feels like in the minds and experiences of our children. It is they who will write tomorrow's history. All children enter their school careers with curious minds. As they move through their school years, they strive to understand the purpose of school and life. Educators, likewise, often grapple with connecting what they are teaching to the children's lives, and even to their own, often asking, "How can I make mathematics come alive, feel exciting, relevant, and meaningful for my children?"

 Check In

As you reflect on that question, challenge yourself to think beyond the norm and imagine a mathematics experience that is inviting for all children. Jot down some words, phrases, or images of what that looks like in the space below.

_____

_____

_____

Let's look at this question through the metaphor of video games. Since their invention, video games have fascinated children and adults alike. Modern video games depict fantasy worlds with endless possibilities offering an exciting alternative reality that helps players explore and problem-solve as a way of making sense of an imaginary world that has challenges and goals within it. Many video games are collaborative, flexible, responsive, involve building and creating, and prize the journey over the destination. There are multiple entry points, multiple destinations, and rarely any prescribed pathways, thus leaning into exploration. These virtual worlds are the kinds of environments that attract children, that children thrive in, and that children *want* to invest their time in.

Unlike video games, the traditional U.S. school system is rigid: educators receive static guidelines (standards) and predetermined pathways (textbooks). They have one single way to win the game (test scores). For students, the school system programs them to log in (start school) and assimilate by following a prescribed pathway (grade levels and disconnected subjects). Stakeholders measure children's performance against their peers and according to specified ideals (more test scores and grades). And if they do everything right, children get to log out (graduate) and become a member of society.

Ironically, the education system also demands that educators prepare children for adulthood and careers in which they must collaborate, problem-solve, communicate, adapt, evolve, create, and imagine. Table 1.1 looks at our current educational system and begins to envision a new system.

This new education system calls on us to acknowledge that as mathematics educators, as social justice educators—no, as *social justice mathematics*

Table 1.1. *Reimagining the Educational System*

What we have is . . .	What we need is . . .
An outdated system of education run like a machine that is designed to produce factory workers.	A system that allows children to explore, solve problems, collaborate, communicate, adapt, evolve, create, and imagine.
A system that takes many inputs (children from various backgrounds) and tries to create a single kind of output (assimilated children), who are then expected to live in a society that was never designed for most of them.	A system that honors all children's unique brilliance and contribution—a system that helps them make sense of and shape the worlds they inhabit so they can build a society that works for all of them.
A system that has produced adults who proudly claim, "I'm not a math person" or "I can't do math."	A system in which adults have confidence in their abilities to use mathematics in their daily life and careers.

educators—we have the responsibility to change the current reality in which, at least in the United States, by the time most children finish their state-mandated, preK-12 schooling—if they even complete it at all—their achievement levels in various educational domains dictate the kind of jobs they may have and how much money they might earn. Becoming a social justice mathematics educator calls on us to envision more for our children and write a new narrative for their future.

The Social Justice Mathematics Teaching Framework

As you envision your future mathematics classroom, you may be asking, *What's social justice got to do with the teaching and learning of mathematics? Do I even want to start down this road?* Take a moment to look at the social justice mathematics teaching framework (Figure 1.1).

This framework consists of five elements: equity, standards-based mathematics instruction, culturally relevant pedagogy, culturally responsive teaching, and teaching mathematics for social justice (TMSJ). In the framework, we intentionally place equity at the outer edge, and the other four elements come together to achieve the ultimate goal of equity:

▶ **Standards-based mathematics instruction** emphasizes the importance of developing—in balance—children's understanding of concepts and procedural fluency, in a discourse-rich setting.

Figure 1.1. *Social Justice Mathematics Teaching Framework*

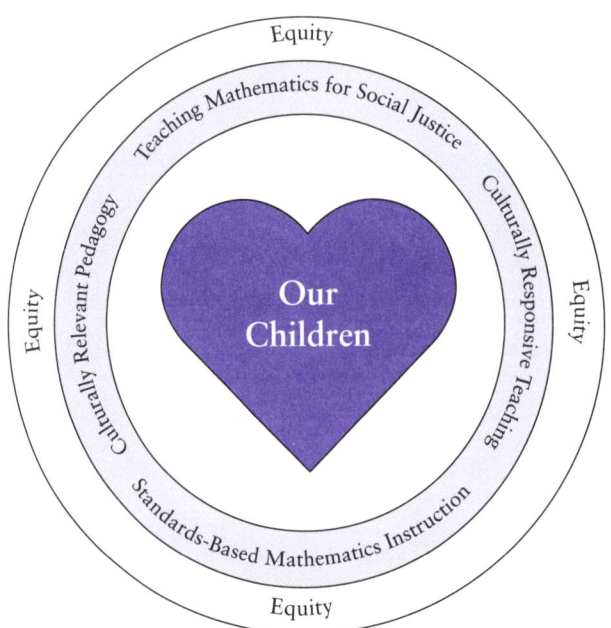

- **Culturally relevant pedagogy** helps children truly understand their own and other cultures and develop a sociopolitical lens.

- **Culturally responsive teaching** incorporates children's lived experiences into their mathematics learning experiences.

- **Teaching mathematics for social justice** operationalizes the other three elements, helping children see how to use mathematics as a catalyst for social change.

It is imperative on the TMSJ journey that one embraces a truly equitable stance to ensure every child has access to relevant and meaningful grade-level mathematics and receives what they need to be academically successful. Each element will be further unpacked in future chapters.

Which "elements" of the framework are you familiar with and which are you ready to learn more about?

CLASSROOM EQUITY

Now that you've started, take a deep breath, smile, and let's look at the question of equity. Often in education, we have treated equity like a bobblehead doll—it looks good, feels good, bounces all around—with no agreed-upon definition to guide the authentic work to create classroom equity. NCTM (2000) states all students, regardless of their personal characteristics, backgrounds, or physical challenges, can learn mathematics when they have access to high-quality mathematics instruction. Equity does not mean that every student should receive identical instruction. Rather, it demands that reasonable and appropriate accommodations be made and appropriately challenging content be included to promote access and attainment for all students (p. 12). Gutiérrez (2012) expands on this definition of equity by adding four specific dimensions: access, achievement, identity, and power, stating that all are necessary for there to be true equity.

- **Access** focuses on the tangible resources available to the children. If a child does not have access to the necessary tools and resources, how do we expect them to be successful? Access includes curriculum materials, technology, high-quality teachers, and an environment conducive to learning.

- **Achievement** is measured by the children's results. Achievement goes beyond a standardized exam. It includes children's engagement in the mathematics experience, their participation in mathematics activities in and beyond the classroom, and their performance on in-class activities and assessments.

- **Identity** gets leveraged by ensuring children have opportunities to be themselves and better themselves while doing mathematics. Children should be able to enter the mathematics experience as their full selves

> Often in education, we have treated equity like a bobblehead doll–it looks good, feels good, bounces all around.

and learn about their classmates during the experience. Mathematics experiences should build upon every child's fund of knowledge they bring into the classroom.

> **Power** takes up the issues of social transformation (Gutiérrez, 2012). In TMSJ, in alignment with Gutiérrez, the power dimension is measured in terms of voice in the classroom; opportunities for mathematics to be used to assess society critically; reimagining the notion of knowledge bearing; and humanizing mathematics.

Through this definition, equity becomes tangible by making each dimension practical, feasible, and measurable over time. In other words, we can operationalize this definition. It is a practical, working definition that anyone can apply to their environment.

In an equitable mathematics classroom, individual factors and experiences, such as those listed below, are recognized, actively included, and celebrated, allowing students to reach their full learning potential.

> Race
> Culture
> Gender
> Religion
> Ethnicity

> Sexual orientation
> Immigration status
> Socioeconomic status
> Learning preferences
> Physical and/or cognitive dis/abilities

In practice, equity means all children receive necessary supports and opportunities in a timely fashion so they can develop their full intellectual, social, emotional, and physical potential to succeed.

In practice, equity means all children receive necessary supports and opportunities in a timely fashion so they can develop their full intellectual, social, emotional, and physical potential to succeed. Equity is different from equality. As described by Chardin and Novak (summarizing Boris Kabanoff), "Equality is the belief that all people have equal value as individuals and should therefore have equal inputs and outcomes. In short, equality is equal distribution where everyone gets the same thing, distributed evenly among them because everyone has the same worth" (Chardin & Novak, 2021, p. 13). But "equity, in comparison, is focused on productivity, or ensuring that everyone has what they need to be successful. This 'fair distributive principle' means that marginalized individuals will need significantly more inputs to have the same, or similar, outputs than individuals with privilege and power" (p. 14). TMSJ is centered on providing such supports and opportunities based on the individual child's needs.

"In order to achieve classroom equity, all adults organizing the learning opportunities—leaders and teachers—must be willing to alter traditional beliefs and practices, adopt an adaptable learning mindset, and implement culturally responsive, engaging materials in order to prioritize success for all students"

(Childs & Davis, 2023). The last part of this statement, "must be willing to alter traditional beliefs and practices," requires us to address school practices that have historically reflected the norms of monolingual, white, middle-class, heteronormative, cisgender, able-bodied, Christian children, and which often exclude children who come from racially and ethnically diverse cultural and linguistic backgrounds, low socioeconomic backgrounds, various religious backgrounds, those who identify as a member of the LGTBQIA+ community, children who identify as having one or more dis/ability, and those who identify with intersections of one or more of these identities.

STANDARDS-BASED MATHEMATICS INSTRUCTION

Standards-based mathematics instruction is grounded in pedagogical principles that emphasize learning for understanding over learning that focuses primarily on procedural fluency and memorization of algorithms and facts. Educators are provided a set of standards that outline the content and mathematical practices and processes all students are expected to achieve. Instruction develops in a discourse-rich environment supported by reasoning and justification (NCTM, 2014).

CULTURALLY RELEVANT PEDAGOGY

Culturally relevant pedagogy (Ladson-Billings, 1995) consists of three tenets.

1. **Academic success:** Children's development of their academic skills.

2. **Cultural competence:** Utilizing a child's culture as a vehicle for learning and providing children opportunities to learn about other children's culture.

3. **Critical consciousness:** Children's development of broader sociopolitical consciousness that enables them to critique the culture norms, values, and institutions that produce and maintain social inequities.

When educators employ culturally relevant pedagogy, they purposefully engage and intentionally lift up the mathematical knowledge bases and problem-solving approaches for the full and diverse range of their students, regardless of the demographics their children present. They actively tap into the cultural backgrounds and social identities of their children and then use those characteristics as conduits for more effective teaching. The ultimate goal is to engage students in critical consciousness, which is an important aspect of K–12 education.

CULTURALLY RESPONSIVE TEACHING

Culturally responsive teaching (Gay, 2018) is defined as "using the cultural characteristics, experiences, and perspectives of ethnically diverse students as conduits for teaching them more effectively (p. 106). In culturally responsive teaching, educators value the importance of children's racial and cultural diversity in the teaching and learning process. In addition to valuing, educators also incorporate these components into the classroom experience. Too often, "children have to attempt to master academic tasks while functioning under cultural conditions unnatural (and often unfamiliar) to them. Children enter schools and are expected to divorce themselves from their cultures and learn according to European-American cultural norms (Gay, 2002, p. 114). As social justice educators, we must collectively work to change this narrative and ensure every child feels seen, valued, and heard.

A key differentiating component between culturally responsive teaching and culturally relevant pedagogy is the critical consciousness tenet. Culturally responsive teaching focuses on intertwining children's cultures with the educational experience, while culturally relevant pedagogy takes it a step further: Now that you have made sense of culture(s) it is time to use that information and take action to impact society. Actions should focus on addressing inequities in society. Thus, the former is focused on solely learning new information and the latter is focused on taking action based upon the newly learned information. The action component leads to TMSJ.

TEACHING MATHEMATICS FOR SOCIAL JUSTICE

TMSJ operationalizes standards-based mathematics instruction, culturally responsive teaching, and culturally relevant pedagogy with an added element of social action. Robert Q. Berry III and colleagues (2020) take the view that "teaching mathematics for social justice (TMSJ) is about teachers emphasizing equitable opportunities for each and every student, as well as developing an orientation toward using mathematics to enact decision-making power" (p. 19).

In TMSJ, the focus is not solely on getting children to just "do something" in the classroom (e.g., perform procedural calculations on a worksheet or apply mathematics to a building project in ways they've been previously shown). Nor does it view children merely as empty vessels to be filled with someone else's knowledge. Rather, the key to TMSJ lies in focusing on the goals of developing children's positive social, cultural, and mathematics identities and, ultimately, classroom equity.

Becoming a Social Justice Mathematics Educator

We take the stance that all mathematics educators are involved in the work of social justice, which includes the following:

1. Cultivating teacher-to-children and children-to-children relationships

2. Questioning systemic structures, policies, and practices that result in inequitable children outcomes

3. Advocating for all children to have access and opportunities to rich mathematical learning experiences

4. Designing mathematical learning to ensure that the goals of developing children's positive social, cultural, and mathematics identities are achieved

5. Taking action that has an impact on a social issue

As social justice mathematics educators, we must be intentional in our work so that children can

1. use their mathematics knowledge and skills for the betterment of their communities and society, and

2. work collaboratively—with persons of all racial or ethnic backgrounds, socioeconomic statuses, and social identities—to conscientiously address the social justice issues that impact themselves and others.

Let's look at where TMSJ comes from.

TEACHING MATHEMATICS FOR SOCIAL JUSTICE IS EQUITY FOCUSED

Berry et al. (2020) further define TMSJ as "much more than the lessons teachers might implement in their classrooms. It is about the relationships they build with and among students; the teaching practices that help them do that; and the goals to develop positive social, cultural, and mathematics identities— as authors, actors, and doers" (p. 23).

Eric Gutstein (2003), professor of mathematics education and curriculum and instruction at the University of Illinois Chicago, for example, identifies the three components of social justice-oriented instruction as

1. helping children develop sociopolitical consciousness,

2. providing children with strengthened senses of agency, and

3. positively highlighting children's diverse social and cultural identities.

Tonya Gau Bartell (2012), associate professor of mathematics education at Michigan State University, posits that mathematics instruction should target three additional goals:

1. teaching children how to apply mathematics to issues of social injustice,

2. helping children develop critical consciousnesses that deepen their knowledge of sociopolitical contexts, and

3. supporting children's involvement in social action.

In Bartell's (2012) view, social justice-oriented mathematics instruction can help children not only understand the world in which they live but also change it. TMSJ also presents a process by which teachers use mathematics to help children understand their roles and places in society and change those roles when social injustice and inequity reign. Bartell argues that by extending the application of mathematics knowledge and skills to real-life matters, children can see mathematics not as something separate from their lives but as an essential aspect of their lives; for example, they may need to determine how much money they could save if they banded together with their neighbors to purchase goods or services in bulk, how many square feet of solar panels they will need to cover the roofs of community-renovated housing and lower the carbon footprint of their homes, or how to assess data from corporate earnings reports or balance a nonprofit organization's budget.

Take a moment to complete Try This: Connections to Mathematics.

 TRY THIS: CONNECTIONS TO MATHEMATICS

Have children identify one personal thing they like to do and one socially impactful thing they are doing or would like to do. For the younger grades, you will possibly need to help them make sense of a socially impactful thing they are doing or would like to do. For example, you may ask them, "What community problem would you like to solve?"

After gathering these two pieces of information, work collaboratively with the children to determine the mathematical connection. This connection should be broad and showcase mathematics beyond just calculations.

Personal	Mathematics Connection (Personal)	Socially Impactful	Mathematics Connection (Socially Impactful)

For example, if a child states they want to reduce violence in their community, this can be quantified through publicly available data. Another example, if a child states they like to sleep as their personal thing (we all know children who will be thinking outside the box), the mathematical connection can simply be the amount of time they like to sleep. Ultimately, the goal of this activity is to show children that mathematics is all around us. The key is to keep the connections simple and use them to launch a discussion.

More recently, the NCSM: Leadership in Mathematics Education and TODOS: Mathematics for All position paper (2016) stresses the importance of mathematics instruction focused on issues of social justice as a way of eliminating the tendency toward deficit views of mathematics learning, reducing the role of mathematics as a "gatekeeper" subject, engaging the sociopolitical turn of mathematics education, and elevating the professional learning of mathematics teachers and leaders. This was followed by the Benjamin Banneker Association (2017) position that mathematics instruction can apply the concept of social justice through three lenses: "about" social justice, "with" social-justice, and "for" social justice. Last, Berry et al. (2020) maintain that social justice-focused mathematics instruction is about teachers emphasizing equitable opportunities for every child and using mathematics to increase children's decision-making power.

The common thread these scholars express is a keen focus on harnessing the power of mathematics as a catalyst for social change. The TMSJ approach shows mathematics educators how to become effective, social justice-oriented teachers who use new ways of thinking about and using mathematics to illustrate and expand children's knowledge about societal challenges. It can guide you to provide your children with mathematics-based skills to help them solve real-life problems resulting from social injustice in their school, communities, society, and the world. Table 1.2 shows the connections among three of the elements by describing each approach's views on academics, learning about

Table 1.2. *Connections Among Culturally Based Pedagogies*

	Culturally Relevant Pedagogy (Ladson-Billings, 1995)	**Culturally Responsive Teaching (Gay, 2002)**	**Teaching Mathematics for Social Justice (Wager & Stinson, 2012)**
Definition	Based on three tenets: 1. Children must experience academic success. 2. Children must develop and maintain cultural competence. 3. Children must develop a critical consciousness through which they challenge the status quo of the current social order.	Using the cultural characteristics, experiences, and perspectives of ethnically diverse children as conduits for teaching them more effectively.	Teaching mathematics for social justice is engaging children in mathematics learning experiences rooted in social issues, with a focus on transforming children's communities.
Views on academics	Children must experience academic success	Standards based	Rooted in rich problem solving
Learning about other cultures	Requires children to become at least biculturally competent	Rooted in educational experiences that integrate various cultures into the classroom	There is a possibility children will learn about other cultures through the social issue, but it is not a requirement
Beyond the classroom	Children learn in a manner to challenge the status quo. Cognitively children will be charged to act; however, it is not a requirement.	Lessons focus on the incorporation of culture into the classroom setting. Learning can extend beyond the classroom but is not a requirement.	Lessons extend beyond the classroom and into children's communities.

Sources: Ladson-Billings (1995), Gay (2002), Wager & Stinson (2012).

other cultures, and its impact beyond the classroom. Note that we did not include standards-based mathematics instruction because we see that as a foundational element of the mathematics classroom, a necessary and foundation for learning.

TEACHING MATHEMATICS FOR SOCIAL JUSTICE IS CHILD-CENTERED

Another aspect of TMSJ is that, beyond being culturally responsive and relevant, the mathematical learning itself must be child-centered. In a child-centered mathematics classroom, children see themselves in the mathematical topic or concept, and the teacher uses cognitively rich thinking tasks. These tasks have multiple entry points and multiple problem-solving pathways that engage all children and guide the children through their own problem-solving process in collaboration with peers to make sense of the mathematics at hand. The teacher is a facilitator who asks questions, provides scaffolding to help children clarify and advance their thinking, and helps them generalize about the mathematics they are coming to understand. This method provides all children in the teacher's classroom with opportunities to engage in the topic by accessing their knowledge base *through their unique identities*. In this way, every child can contribute to the mathematics classroom. Every child is encouraged to bring what they know mathematically—based on their unique cultural understandings—into what you, as a future social justice mathematics educator, will come to see and convey as the "whole universe of learning." TMSJ recognizes that every civilization and people of diverse cultures and worldviews have contributed to the development of mathematics and that children bring their own wealth of knowledge and interesting ways of thinking and sensemaking to the classroom. Thus, the wealth of mathematics knowledge, skills, and practices that diverse groups of children embody should be actively celebrated and shared, respectfully and constructively, to enhance all children's mathematics learning and understanding.

All children, not just melanated children, need to understand the history of mathematics and how it arose from many cultures. They should understand that although mathematics initially arose from a need to count and record numbers, people engaged in mathematics while going about their practical lives beyond just the modern-day focus on algorithms and solution-getting (Joseph, 2011). Mathematics was used to make sense of the world, from understanding time to travel guidance to spiritual understandings and connections. This understanding is important for children because as naturally inquisitive people, they can see how others used math to make sense of their lives. Joseph (2011) provides countless examples of the multicultural roots of non-European mathematics, which will help today's racially and ethnically diverse children better understand their place in history and how their ancestors contributed to mathematics.

For example, in Africa, the Ishango bone, an engraved bone more than 10,000 years old, is believed to have been used to count, play games, and engage in lunar observations, among other uses. The ancient Chinese used mathematics for fraction operations, quadratics, trigonometry, and other uses. Ancient Indian civilizations discovered the sine functions and used mathematics for astronomy and navigation, among other uses. Islamic contributions to mathematics included geographic uses, analysis of property relations, and distribution of inheritances. These are just a few instances of ancient civilizations' uses of and contributions to mathematics. The more we, as educators, can de-center the "Greek" lens when exploring the history of mathematics, the more children can have an opportunity to see themselves in mathematics and the better white-identifying children will understand the essential and rich contributions of others.

TMSJ also supports the development of the whole child by preparing them with the knowledge, skills, attitudes, and values to thrive in their classroom, school, community, and beyond. Table 1.3 lists the competencies and skills from three resources that inform this work: the social-emotional well-being competencies outlined by CASEL (n.d.), the P21 (2019) *Framework for 21st Century Learning*, and the OECD (2019) *Learning Compass 2030* (see Chapter 6 for additional connections).

> The more we as educators can de-center the "Greek" lens when exploring the history of mathematics, the more children can have an opportunity to see themselves in mathematics and the better white-identifying children will understand the essential and rich contributions of others.

Table 1.3. *TMSJ Supports Whole Child*

Social & Emotional Learning	21st Century Learning	Learning Compass 2030
• self-awareness • self-management • responsible decision making • relationship skills • social awareness	*Learning and Innovation Skills* • creativity and innovation • critical thinking and problem solving • communication and collaboration *Life and Career Skills* • flexibility and adaptability • initiative and self-direction • social and cross-cultural skills • productivity and accountability	*Core Foundations* • cognitive foundations (literacy and numeracy, upon which digital literacy and data literacy can be built) • health foundations (including physical and mental health and well-being) • social and emotional foundations (moral and ethics) *Transformative Competencies* • creating new value • reconciling tensions and dilemmas • taking responsibility

Sources: CASEL (n.d.), P21 (2019), OECD (2019).

TEACHING MATHEMATICS FOR SOCIAL JUSTICE IS ACTION-ORIENTED

TMSJ aims for action as its ultimate goal. It commences well before planning to involve children in a social justice mathematics lesson. The classroom setting holds significant importance because children are expected to learn alongside classmates from various backgrounds, races, ethnicities, and cultures, thereby gaining exposure to diverse perspectives. The initial steps involve learning about oneself, exploring identities, biases, and beliefs (refer to Chapter 2) and understanding the children, their identities, strengths, and interests (refer to Chapter 3). These actions are pivotal in establishing a mathematical learning environment (see Chapter 4) that engages all children in the mathematics classroom (see Chapter 5) and fosters a sense of belonging and of being valued. This process demands your commitment to learning, potentially prompting adjustments to instructional practices and the acquisition of new practices.

As you implement social justice mathematics lessons, you'll have the chance to gather input from your children and families regarding current community issues (see Chapter 6). Collaboration with families, community members, and colleagues in the planning process might result in their involvement in the actual lesson (see Chapter 9). Following or even during the lesson, children should be provided with opportunities to explore ways to expand and apply their classroom learning to actions they can take within their school, community, and beyond (see Chapter 8). Consequently, every facet of TMSJ entails some form of action.

To hear more from the authors about the social justice mathematics teaching framework, listen to this **conversation with Dr. Childs and Dr. Staley.**

qrs.ly/xmfful9

To read a QR code, you must have a smartphone or tablet with a camera. We recommend that you download a QR code reader app that is made specifically for your phone or tablet brand.

Summary

In this chapter, we introduced a social justice mathematics teaching framework that consists of five elements: equity, standards-based mathematics instruction, culturally relevant pedagogy, culturally responsive teaching, and teaching math for social justice (TMSJ). After examining each of the elements, we concluded the chapter with a discussion of the child-centered nature of TMSJ and also discussed how the various aspects of TMSJ are action oriented.

As you set off and work your way through this journey to becoming a social justice mathematics educator, we want to provide you with the right tools. One of those is the TMSJ Action Plan (Appendix A). You can access this on the book's companion website and use it with each chapter of the book. We have provided space for you to record your key takeaways from each of the chapters and encourage you to identify possible next steps to support you as you continue through each stage of your social justice mathematics teaching journey.

online resources The TMSJ Self-Assessment is available for download at https://qrs.ly/wbfixtr

REFLECT

- How do the elements of the equity-based mathematics teaching framework come together in TMSJ? Which component(s) do you need to note for further exploration?

- Why does teaching mathematics for social justice (TMSJ) matter to you?

_____

_____

_____

_____

_____

_____

_____

_____

_____

_____

_____

_____

_____

ACT

1. Add your key takeaways and next steps to your TMSJ Action Plan.

2. Complete the following TMSJ Self-Assessment:

_____

_____

_____

_____

_____

_____

_____

_____

_____

_____

_____

_____

- What are the first three thoughts that come to mind when you think of social injustice?
- Write down three words or phrases that describe your own mathematics teaching style.
- What emotions come to mind when you think about the intersection of mathematics and social justice? What excites you? What worries you?
- What is the biggest social injustice issue in the community in which you teach? Why?
- Now that you have responded to the preceding questions, use the following chart to reflect on where you are.

	Beginner	Intermediate	Advanced
Self-Reflection	←————————————————→		
Children Under Your Purview	←————————————————→		
Classroom Environment	←————————————————→		

online resources ⬏ Available for download at https://qrs.ly/wbfixtr

Where to Next?

If you, as a mathematics educator at any grade level, realize that *all* children should be seen and heard in your classroom, then TMSJ is for you. If you believe that they can bring, even into a mathematics environment, their authentic, culturally informed mathematics knowledge and apply it effectively and creatively to real-world problem solving, then TMSJ is for you.

If you want to present your children with rich, high cognitive–demand mathematics tasks and lessons that have meaning beyond the classroom setting, then TMSJ is for you. Again, TMSJ will help you propel your children toward success in many real-world mathematics applications.

Last, if you are tired of culturally monolithic, monochromatic mathematics classrooms that fail to buzz with the excitement of authentic discourse and interaction and that are no way near sites of continuous feedback that can help you escalate children's learning, then you should start your TMSJ journey *now*! The best way to do that is to head to Chapter 2, which will help you engage in the critical self-reflection that is a must-do before shifting your instruction to a TMSJ orientation.

MIRRORS: UNDERSTANDING YOURSELF

2

Now that we have gained a foundational understanding of what teaching mathematics for social justice (TMSJ) means, it is time to set you off on what we hope is a transformational journey. We recognize that every educator will enter into this work at a different place, so the key is for you to focus on you and where you are today. This chapter is meant to help you do that. In this chapter you will

▶ engage in a personal reflection about your various identities as a human, an educator, and a teacher of mathematics;

▶ explore the connections between your identities, beliefs, and biases; and

▶ gain an understanding of how decentering your own identity and whiteness in the classroom will make room for the identities of the children under your purview.

My Identities Matter

People have all kinds of identities—some outward-facing and some inward-facing. Your identity markers include your race and ethnicity; gender; sexual identity; age; religion or nonreligion; native language; socioeconomic status; education; profession; relationships and positions with other people; and physical, mental, and emotional abilities. Some identity markers are given to us and are fixed, while others are changeable and have varying impacts on our lives. Take a moment and consider, "What identities describe you when you are at work?" In every space you enter, you bring a lived experience that may be similar to or different from the experiences of the children in your classroom. Understand that the value you place on specific identities can change over time. In addition, you may have some identities that you do not consciously consider because they may not have a regular impact on your daily life. Take some time to complete the Try This: Reflecting on My Identity Markers as you unpack your identity markers, how they impact your life, and how they impact you as an educator. This can be completed individually or as a part of a group activity. Allow yourself ample time to complete the activity and reflection questions. If you are working alone, fully explain your thoughts and deeply reflect upon your responses. If you are working with a group, carefully respond to the reflection questions according to your personal comfort level. You are not under any obligation to share pieces of your identity that you are not comfortable with.

 TRY THIS: REFLECTING ON MY IDENTITY MARKERS

Mark the top five identity markers that impact your role as an educator, then complete the reflection questions.

Age	Religion
Gender	Immigrant Status
Race	Physical Appearance
Ethnicity	Sexual Orientation
Education	Family Status
Geographical Location	Language
Nationality	Physical Ability (able-bodiedness)
Marital Status	Socioeconomic Status

Reflection Questions

1. Which three of the five identity markers are your most important identities? Why?

2. Which ones do you think others typically notice about you? Why?

3. Which three impact your role as an educator the most? Why?

4. When in your educational environment are there identities you prefer to minimize? Why?

5. What feelings arose as you completed this reflection? As you shared your reflection?

_____
_____
_____
_____
_____
_____
_____
_____
_____
_____
_____

Source: Adapted from Aguilar (2021).

Now that you have reflected upon your identity markers, you are probably still asking yourself, "Why is it important to critically reflect on one's identities? How does this impact my role as an educator?" To be a good educator, this activity does not necessarily matter; an educator's job is to teach children specific content and then go home. However, as an educator committed to TMSJ, your work goes beyond just teaching children content. It expands to teaching children about life and how mathematics can be used to improve their lives and communities.

Since the beginning of time, mathematics has played a role in every civilization (Joseph, 2011), and TMSJ provides the space to continue this legacy. This starts with each of us truly understanding who we are and how we enter this work. Understanding who you are through your identity markers helps you better understand how you show up in different spaces, including in the educational setting, and more importantly, how your instructional practices are formed. Let's explore the intrinsic values of one's identities as we look back at Try This: Reflecting on My Identity Markers. Think about how much these identities mean to you and how they shape your life. Reflect on how often you consciously consider these identities and the role they may or may not play when you are in your educational environment.

> As an educator committed to TMSJ, your work goes beyond just teaching children content. It expands to teaching children about life and how mathematics can be used to improve their lives and communities.

Check In

Now, imagine that tomorrow, when you enter your educational setting, you have to leave the three identities you care about the most at the entrance.

What feelings are arising? Do you believe you will now be successful in the educational setting? Would you still want to enter?

Picture yourself as a child entering the classroom or school. Before crossing the threshold, what identities must the child leave at the door to assimilate into the setting and survive? Every day there are children who enter our schools and classrooms and they know some of their identities are not valued in the space. For example, there may be a child who identifies as LGBTQIA+, and by not using the child's preferred pronouns, the school has shown the child they are not valued and won't be seen. The child is dehumanized, not because of anything they have done, but for simply existing. Now consider the child whose identities do not match many of the identities within the school or classroom. It is one thing when identities do not match; it is another when, throughout the day, adults organizing activities such as musical events, curriculum projects, and other extracurricular activities never align with the child's identities while, in contrast, other children's identities are repeatedly validated.

We must remember how we felt imagining that, as adults, we would have to leave our identities at the door and consider how our children are—intentionally and unintentionally—subjected to educational settings where they are forced to do the same and must adapt to the setting. We challenge you as a reader to consider your setting and ask,

> ▶ What is being done to ensure every child can authentically be themselves?

> ▶ What is being done to eradicate any barriers to a child being their full self?

> ▶ What am I doing to ensure the educational setting is inclusive?

THE VISUAL IDENTITY MARKERS

Let's further explore your identity markers and how they have impacted your career and your role as a mathematics educator. Do you feel your identities have enhanced or hindered your career opportunities? Why? Now consider your race, ethnicity, culture, and gender identities (see Figure 2.1). How have these identity markers impacted your career? How do they show up in your mathematics classroom setting?

We focus on these four identity markers because they play a keen conscious and unconscious role in the mathematics classroom. They are the identity markers that are observed when someone enters a room, and assumptions are naturally made based on what people see. We say "assumptions" because one does not truly know a person's identities unless they tell you. However, these four identities by which folx are generally viewed and categorized shape society. Race and ethnicity are deeply rooted identity constructs that have been consistently used as a measuring stick in society and have become part of everyone's social conditioning. Together with culture and gender, these identity constructs are

Figure 2.1 *Your Visual Identity Markers*

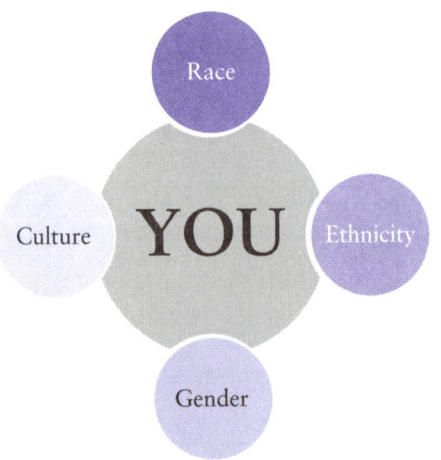

deeply embedded in the fabric of our culture, and they play a major role in the beliefs and biases that are formed about you and your children.

We have to be honest with ourselves and acknowledge that we live in a gender-oriented, racialized, and ethnically based society. Nearly every form we fill out asks us about our gender, race, or ethnicity. Personal data is tied to our race and ethnicity. Where we live, what schools we attend, where we grocery shop, how we interact with people, what languages we speak, what food we eat, and what customs we cherish are all often tied to race and ethnicity. We live with the legacy of policies and laws based on people's race and ethnicity. Student achievement data, course placement and acceptance, and graduation rates are all bound together with data about race and ethnicity. Our diversity is beautiful and is a strength, and yet for many, it can present many barriers. Many of us, regardless of race or ethnicity, have never deeply investigated what this means at the individual level and how these two specific identity markers play a major role not only in our personal lives but also our professional lives and in relation to the children in our classrooms.

Check In

During the school day, how often do you think about the race, ethnicity, gender, or culture of

- yourself?
- your colleagues?
- children you teach?
- contextual settings for lesson tasks and experiences?

MY IDENTITIES SHAPE MY BELIEFS

Your personal identity markers are vital to reflect on as you start to think about your practices because they have been at the core of shaping who you are as a person and educator through your background and lived experiences, inside and outside of educational settings. It is important to ask yourself, "How do my identities impact my teaching and the children in my care?" To understand how your identities shape your teaching practices, we start by considering the beliefs we hold about your children as people; thinkers, doers, and creators of mathematics; and ultimately learners worthy of engaging mathematical experiences. Take some time to complete

TRY THIS: IDENTITIES AND BELIEFS

1. List the top five identity markers you identified in Try This: Reflecting on My Identity Markers.

2. For each identity marker, consider how it shapes your belief about your children as doers and learners of mathematics. Record your reflections in the chart below.

My Identity Markers	Beliefs About My Children

online resources Available for download at https://qrs.ly/wbfixtr

Try This: Identities and Beliefs as you reflect on how your identities shape your beliefs.

As you reflect on your identities and how they shape your beliefs about your children, we invite you to begin to think about how this shows up in your practices: mindset, attitudes, and behaviors. You might notice that your beliefs lead to practices that are driven by your "likes" and "dislikes." In other words, you are beginning to identify your biases about your children as learners and doers of mathematics.

Identifying My Biases and Beginning the Work to Address Them

> The Plague and power of bias are too consequential to let them go unacknowledged and unchecked. They can affect us in surprising ways. (Eberhardt, 2019, p. 30)

We now turn your attention to unpacking your biases—those you are aware of and those that you are not. "We have a bias when, rather than being neutral, we have a preference for (or aversion to) a person or group of people" (Colorado Department of Education, n.d.). Figure 2.2 provides a visual of the underlying concepts of bias.

Figure 2.2 *Underlying Concepts of Bias*

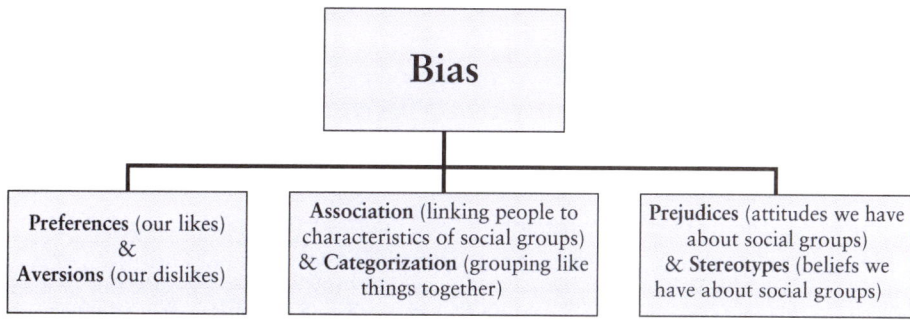

Source: Adapted from Eberhardt (2019).

Because your biases condition how you look at and relate to your children, it is vital that you understand them, acknowledge the consequences of your checked and unchecked biases, and mitigate those biases that need to be addressed. The biases you carry can affect

➤ *what you perceive about your children*—the way you view them and their families,

➤ *how you think about them*—beliefs about their abilities to learn and do math, and

➤ *the actions you take*—the mathematics environment you create, the learning experiences you design, and the relationships you develop with your children.

We also have unconscious biases (or implicit biases) that associate stereotypes or attitudes toward categories of people without conscious awareness, which can result in actions and decisions that are at odds with one's conscious beliefs about fairness and equality (Osta & Vasquez, n.d.).

There are two questions that you may be asking yourself: "How can I uncover my own biases?" and "What implicit biases do I have that must be addressed?" Answering these questions requires you to take a critical look at yourself and through the "windows" (see Chapter 3) at your children to identify

1. your likes and dislikes about who you teach, where you teach, and what you teach;

2. the ways children are grouped or labeled in your school and school community;

3. your beliefs and attitudes about your children, school, school community, and teaching and learning of mathematics.

Take a moment to complete the Try This: My Biases as you start to identify your biases.

TRY THIS: MY BIASES

- Review your responses to Try This: Identities and Beliefs and select one identity marker and its belief statements to record in the chart below. (Note: In the Reflection and Action at the end of this chapter, we invite you to complete this exercise for the other four identity markers and beliefs.)

- Capture your thoughts for each prompt as you consider the impact of your biases on who you teach, where you teach, and what you teach.

- Review your responses and identify (underline, circle, or highlight) where you are not able to be "neutral" because you have a strong preference or aversion. These are your biases!

My Identity Marker	Beliefs About My Children
*What are your preferences (likes) and **aversions** (dislikes) about . . .*	
Who you teach (children & families)	
Where you teach (school and school community)	
What you teach (grade levels, mathematics courses/content)	
*In what ways do you **associate** (group) or **categorize** (label) . . .*	
Who you teach (children & families)	
Where you teach (school and school community)	
What you teach (grade levels, mathematics courses/content)	
*What are you stereotypes (beliefs) and **prejudices** (attitudes) . . .*	
Who you teach (children & families)	
Where you teach (school and school community)	
What you teach (grade levels, mathematics courses/content)	

online resources Available for download at https://qrs.ly/wbfixtr

Now that you have identified your biases, seek out a critical friend to discuss your responses with and ask them to provide you with additional feedback on things you might have missed—your "blind spots"—when you looked in the mirror. (Chapter 9 goes into further detail about the importance of relationships with colleagues as you continue your TMSJ journey.) It is often difficult to identify your "blind spots," also known as implicit biases, as these are the areas that don't show up in our conscious knowledge. A critical friend can provide insights about how you talk about your children, classroom, school, or school community; your interactions with children in the classroom and school; your daily preparation and planning for mathematics class; and the instructional approaches and tools used in your mathematics classroom, just to name a few areas.

Hopefully you now have identified a few biases and are ready to begin the work of checking those that show up in your mathematics classroom and especially in your interactions and relationships with your children. Table 2.1 provides four areas for self-reflection with prompts around identity, language, relationships, and behavior. We also include Actions to Consider and What It Looks Like to assist you in identifying any biases that need to be checked.

Table 2.1   *"Checking" My Biases*

Notice & Wonderings	Actions to Consider	What It Looks Like in the Mathematics Classroom
My Identity What connections can you make between your identity and your biases?	• Acknowledge the biases that are connected to your identity and consider how they might show up in the mathematics classroom. • Select one to address/be more conscious of and check as needed.	Monitor use of a preferred instructional approach (i.e., avoid using procedural approach if not aligned to the intention of the standards for a lesson).
My Language How do your biases show up in the words you use when speaking about your children, families, classes, school, or school community?	• Identify your use of deficit language—words, terms, labels that focus on lack, loss, or a negative perspective—when describing your children's mathematics ability, effort, or performance, and commit to shifting your speech to an assets- or strengths-based approach. • Commit to eliminating deficit language. Become a strength finder by seeking out at least one asset in each of your children.	Reframe language and labels about children, i.e. shift from "struggling children" to "children needing additional support to access grade-level mathematics."

(Continued)

Table 2.1   "Checking" My Biases (Continued)

Notice & Wonderings	Actions to Consider	What It Looks Like in the Mathematics Classroom
My Relationships Who do you invite into your space?	• Learn about your children's backgrounds, interests, funds of knowledge, and strengths. • Read Chapter 3, Windows: Understanding Your Children	Have your children complete a "Math Autobiography" or survey where you gather information about their interests, strengths, and needs inside and outside the mathematics class. Review children's responses and be intentional about using the information to build connections to mathematics.
My Responses How do your biases manifest in your interactions with the children you teach?	• Monitor your classroom interactions (i.e., responding to questions, providing feedback, redirection statements).	Respond/provide feedback to all children that shows that they are valued as thinkers, doers, and learners of mathematics.

As you look in the mirror and reflect on your identities and do the work of identifying your biases and implicit biases, we now turn your attention to whiteness—how it might show up, the role you and other educators might play (often unconsciously) in centering it, and the importance of decentering whiteness to establish a TMSJ environment.

How Does Whiteness Show Up in the Mathematics Classroom?

In an effort to be relatively generic, mathematics problems often center on an "average" middle-class context, describe students who likely identify as white and have European-derived names, and describe generic activities many children may not be exposed to or interested in. This generalizing and whitewashing results in many students never seeing their culture, their race, or their community represented in the mathematics they are learning. It doesn't help them see how math is for them. It also does a disservice to white children who never have a chance to learn the deep, rich, and complex history and origins of mathematics, and it reinforces whiteness as the norm. Some curriculum publishers try to shift this by changing the names of characters in story problems to sound more "ethnic" or changing the object of a problem, such as changing from dividing up sandwiches or pizza to dividing up burritos to investigate fractions (which is a problem because no one can eat the middle third of a burrito). Though well-intentioned, this, in many ways, actually does *more* of a disservice because it often reinforces stereotypes and lacks any depth, authenticity, or meaning to the children in *your* classroom. Similarly, many of the generic posters or other images teachers put on walls depict math as the domain of old, white, men. Many teachers are engaging in a more active effort to disrupt these norms by changing the contexts in their textbook problems to be more representative of their students; some teachers will put posters on their walls of mathematicians who are Black, Latin@, women, or queer to help broaden the ideas around what math is and who it is for. Take a moment to complete the Check In as you reflect on your school and classroom settings.

 ## Check In

What do the artifacts in your classroom look like? If you walk down the hallways or passages of a school in your area, what do you see? Where might your own biases affect how mathematics gets represented in your classroom?

DO I CONSCIOUSLY OR UNCONSCIOUSLY CENTER WHITENESS IN MY MATHEMATICS CLASSROOM?

When teaching mathematics for social justice, educators must ask themselves, "Do I consciously or unconsciously center whiteness in my mathematics classroom?" Before answering this question, we offer a definition of whiteness. DiAngelo (2011) defines whiteness not as just the identity components of race and ethnicity but as a multitude of processes and practices that include basic rights, values, beliefs, perspectives, and experiences and that center and consistently benefit those who identify as white. Now let's look critically at our own mathematics classroom to determine whether and how we are consciously or unconsciously centering whiteness. There are many ways in which multiple identities are often elevated or stifled in our classrooms, but one of the most critical ones we must first address is the dominance of whiteness.

Addressing whiteness is an ongoing journey, and success doesn't mean tackling all aspects perpetuating whiteness in your classroom at once. Instead, it's about starting urgently and persistently addressing each indicator of whiteness one step at a time, in an ongoing process. This leads us to the question, "What happens if one *does* or *does not* address whiteness in their mathematics classroom?"

▶ **If one does not address** whiteness in the mathematics classroom, inequities in education and in society will remain a perpetual cycle. Historically excluded children will continue to experience subpar mathematics learning experiences.

When teaching mathematics for social justice, educators must ask themselves, "Do I consciously or unconsciously center whiteness in my mathematics classroom?"

> ▶ **If one addresses** whiteness in the mathematics classroom, then the environment begins to become more inclusive, and children see themselves within the experience. This leads to an increase in children's engagement in mathematics, which in return will improve their academic achievement.

WHY IS IT IMPORTANT TO DECENTER WHITENESS?

TMSJ aims to foster equity in math classrooms and inspire children to pursue social justice in their schools and communities.

Traditional U.S. mathematics education has long reflected the dominant culture of white, European backgrounds, consciously or unconsciously shaping educators' approaches (Cintron et al., 2001). TMSJ aims to foster equity in math classrooms and inspire children to pursue social justice in their schools and communities. The approach doesn't solely focus on race or ethnicity but addresses and rebalances power dynamics related to race. It strives for inclusive learning experiences that embrace diverse identities (Ladson-Billings, 2021), crucially decentering whiteness. TMSJ aligns with culturally relevant and responsive teaching, aiming to create equitable math classrooms that catalyze positive change.

Peggy McIntosh (1998), a feminist and anti-racism activist, defines whiteness as beyond racial identity, emphasizing its historical elevation of white people and their societal advantages over nonwhite individuals globally. Over centuries, racism and white supremacist ideologies have served to elevate white people—socially, economically, politically, culturally, and in other ways—over nonwhite people around the globe. In the United States and elsewhere, whiteness allows white children and adults to have their lives shaped, consciously or unconsciously, via unearned privilege or protections solely because of their skin color. TMSJ deliberately displaces whiteness from its customary role in math education, exposing the unjust advantage of skin color in the classroom while offering fair opportunities for learners from various cultures and identities to engage with mathematics in their daily lives and communities.

TMSJ integrates culturally informed math principles and practices from nonwhite cultures into everyday lessons, laying the groundwork for teaching math through a lens of social justice.

In the U.S., whiteness historically dominates, akin to baseball and apple pie, but decentering it allows educators to understand nonwhite children's identities better and create more inclusive classroom experiences. TMSJ integrates culturally informed math principles and practices from nonwhite cultures into everyday lessons, laying the groundwork for teaching math through a lens of social justice.

HOW CAN I DECENTER WHITENESS IN MY CLASSROOM AND MAKE ROOM FOR OTHER IDENTITIES?

The process of decentering whiteness starts by examining how much your school and classroom focus on white perspectives. Table 2.2 offers a set of questions and indicators to guide this evaluation. The downloadable tool helps assess the degree of white-centric content in images, math tasks, and behaviors, providing a way to capture this information.

Table 2.2 *Creating and Cultivating Inclusive Classrooms*

Decentering Questions	Indicators or "Look-Fors" That Decenter Dominant Culture in Your Mathematics Classroom
Does your curriculum focus on a singular problem-solving approach to a mathematics task?	**Multiple Solution Methods:** Emphasis on more than one specific problem-solving method or strategy across various mathematical tasks and topics. **Flexibility in Instruction:** Teacher provides more than one method or approach of problem-solving and provides opportunities for children to experiment with different approaches. **Divergent Thinking Encouraged:** Emphasis on divergent thinking or creativity in problem-solving, encourages children to explore multiple pathways to solutions. **Varying Range of Examples:** Examples provided in lessons show varying methods or approaches, offering exposure to diverse problem-solving techniques.
Are children primarily provided problems or tasks to solve individually rather than collectively? Do your instructional practices and grading policies encourage competition or cooperation?	**Independent and Group Task Distribution:** Individual and group tasks or problems are incorporated into the lesson plan. **Group Discussions and Problem-Solving:** Children are encouraged to discuss mathematical problems collaboratively with peers during class time. **Cooperative Learning Structures:** Activities or structures specifically designed to encourage teamwork, problem-solving discussions, or collaborative problem-solving. **Values Collectivism:** Classroom norms encourage collaborative or cooperative learning, fostering teamwork, collective thinking, and collective problem-solving. **Peer Interaction:** Structured time dedicated to peer-to-peer interaction or sharing of strategies when solving math problems. Children are encouraged to critique the reasoning of others. **Group Accountability:** Assessments evaluate group performance to foster collaborative problem-solving skills.
Do the images (people, places, things) represent various cultures?	**Diversity in Representation:** Visuals authentically showcase people, places, or things from a variety of cultures or identities. **Historical or Global Perspectives:** Images focus on historical events, mathematical contributions, or settings from a diverse cultural lens, highlighting broader global or historical contexts. **Cultural Context:** Images exhibit context or explanations about different cultural practices, traditions, or contributions, exposing children to diverse cultural perspectives. **Multicultural Context:** Images display cultural diversity, with a portrayal of cultural groups reflecting the multicultural reality of the classroom, school, or community. **Void of Stereotypical Depictions:** Images do not perpetuate stereotypes or clichés about specific cultural groups, presenting them in a narrow or limiting manner.
Are students expected to speak in English a majority of class instead of their native language?	**Multilingual Learning Materials:** Math-related materials, including textbooks, worksheets, support materials, or instructional resources, are available in children's native language to aid comprehension or reinforce mathematical concepts. **Language Flexibility:** Children are encouraged to use their native language to discuss mathematical concepts or problem-solving strategies when grouped with peers who speak the same language. **Children Self-Choice:** Children express mathematical ideas or ask questions in their native language. **Multilingual Collaboration and Group Discussions:** Collaborative problem-solving activities or peer interactions related to math problems incorporate the use of multiple languages, allowing children to leverage their linguistic strengths.

(Continued)

Table 2.2   *Creating and Cultivating Inclusive Classrooms (Continued)*

Decentering Questions	Indicators or "Look-Fors" That Decenter Dominant Culture in Your Mathematics Classroom
Do classroom norms and rules focus on behavior compliance?	**Behavior Expectations:** Classroom norms and expectations consider diverse cultural practices or expressions, prioritizing children's active engagement and collaborative discussions. **Culturally Responsive Practices:** Classrooms include practices or norms that honor and incorporate diverse cultural values, traditions, or ways of knowing into the learning environment. **"Check" Cultural Bias in Discipline:** Disciplinary actions or consequences that disproportionately affect children from cultural backgrounds different from the dominant culture are monitored and minimized. **Flexibility in Expression:** Classroom norms are created that encourage the expression of diverse cultural identities, language, or customs.

[online resources] Available for download at https://qrs.ly/wbfixtr

A culture in which whiteness is seen as the default fails the majority of children in our schools and robs all of them of potentially having richer and fuller educational experiences. This is why we must individually and collectively work to decenter whiteness. Note the keyword is *decenter*, not *eliminate*. To do this, use the questions in the Check In below to self-reflect.

Check In

- Do the values, beliefs, perspectives, and experiences I elevate in the mathematics classroom tend to reflect and benefit those children who identify as white?

- Do I consciously or unconsciously establish norms or rules that disproportionately penalize nonwhite children?

If you answered yes to either of the questions, you have consciously and/or unconsciously centered whiteness in your mathematics environment. Each question signifies an issue within mathematics education that needs to be critically evaluated to begin the process of creating an equitable mathematics classroom that serves as the foundation for TMSJ.

In Chapter 3 we will take a closer look at the identities of the children in the classroom, whose identities are already centered, and what it will take to decenter identities—especially ones centered in whiteness—so that every child can be seen, valued, and heard in the classroom.

Summary

In this chapter, we encouraged you to make time to focus on yourself by first looking at your identity markers and how they might shape your beliefs. Next, you examined bias—likes/dislikes, categorizations/associations, and prejudices/stereotypes—with the goal of identifying how they show up in your language, relationships, and behaviors. We concluded the chapter with a discussion of whiteness in the mathematics classroom—how it shows up and the importance of decentering it.

To hear more from the authors about their reflections on understanding yourself, listen to this **conversation with Dr. Childs and Dr. Staley.**
qrs.ly/thffuln

 REFLECT

We were intentional in placing this chapter, Mirrors: Understanding Yourself, at the beginning of the book as TMSJ does not happen in the classroom unless you make this first decision to begin the journey of self-reflection. Take a moment to reflect on and answer the following questions:

- What's the connection between my identities, beliefs, and practices?
- How do my identities impact my teaching and the children in my care?
- What biases do I need to address?
- What can I do to decenter whiteness?

_____

_____

_____

_____

_____

_____

_____

_____

_____

_____

_____

_____

_____

_____

_____

_____

_____

 **ACT**

1. Add your key takeaways and next steps to your TMSJ Action Plan.

2. Complete the Try This: My Biases for your remaining four identity markers and identify your biases that need to be monitored.

_____

_____

_____

_____

_____

_____

_____

_____

_____

_____

My Biases

- Review your responses to Try This: Identities and Beliefs and select one identity marker and its belief statements and record them in the chart below.

- Capture your thoughts for each prompt as you consider the impact of your biases on who you teach, where you teach, and what you teach.

- Review your responses and identify (underline, circle, or highlight) where you are not able to be "neutral" because you have a strong preference or aversion. These are your biases!

My Identity Marker	Beliefs About My Children
What are your preferences (likes) and **aversions** (dislikes) about . . .	
Who you teach (children & families)	
Where you teach (school and school community)	

My Identity Marker	Beliefs About My Children
What you teach (grade levels, mathematics courses/ content)	
*In what ways do you associate (group) or **categorize** (label) . . .*	
Who you teach (children & families)	
Where you teach (school and school community)	
What you teach (grade levels, mathematics courses/ content)	
*What are you stereotypes (beliefs) and **prejudices** (attitudes) . . .*	
Who you teach (children & families)	
Where you teach (school and school community)	
What you teach (grade levels, mathematics courses/ content)	

online resources ☞ Available for download at https://qrs.ly/wbfixtr

Where to Next?

The question we are most often asked when presenting on TMSJ is "When can I start incorporating TMSJ into my children's classroom experiences?" Our answer: You can start immediately! Incorporating TMSJ into your mathematics lesson plans is very much like learning to ride a bicycle. There is no perfect time to start, but the process can be very challenging at first. Finding your balance and learning to use the gears (if any) and brakes may be hard. You may encounter several bumps in the road, but with time and practice, you will see improvements and both you and your children will "get the hang of it." Once you get started, you will continually progress and improve. You will probably find yourself trying new, previously unimaginable things in your mathematics classroom and realize success after success as your children react and respond to TMSJ instruction. The next step in your journey is found in Chapter 3, where you will look through some windows to consider the children in your own classrooms.

3

WINDOWS: UNDERSTANDING YOUR CHILDREN

Although the work of teaching mathematics for social justice (TMSJ) starts with self-reflection, that is only the beginning of the work. Equally important is getting to know the children in your classroom; understanding what they value; what drives them personally and academically; what cultures and customs inform their identities; what their strengths are; and what interests, joys, and struggles they face as individual young humans. This chapter guides you in the critical work of learning about who your children are and what the TMSJ journey can look like for them. In this chapter, you will

> understand the backgrounds, funds of knowledge, and identities of the children in your classroom;

> examine how students' identities play a factor in your classroom;

> consider ways to employ a strengths-based lens to the children in your mathematics classroom; and

> explore how you can build relationships, get to know your children, and uncover and build on their strengths and funds of knowledge.

Social justice mathematics educators always have a keen focus on centering children. As one seeks to center them, it's important to first have a solid understanding of who they are and the identities they bring into the classroom.

My Children's Identities Matter

Social justice mathematics educators always have a keen focus on centering children. As one seeks to center them, it's important to first have a solid understanding of who they are and the identities they bring into the classroom. This must go beyond a superficial and visual understanding to a learned and authentic understanding of who they are. This section answers the following questions:

> Who is in my classroom?

> What identities do my children bring into the classroom?

> How does race play a particular role in my classroom?

WHO IS IN MY CLASSROOM?

As mentioned, it is important to know yourself, your identities, and your biases, but it's even more important to understand your children's interests and background—personal, family, cultural, and social experiences that shape their identity and understanding of the world. Whether you teach 20 or 120 children a day, each child walking through your door brings their lived experiences and developing identity. We must grasp that although some identities are static and inherent from birth, others evolve over time due to societal views or the child's own development. Children, like us, have identities that can change, and they deserve the space and freedom to grow. This growth isn't solely long-term, spanning from elementary to high school, but also occurs within a single school year.

Children develop a set of identities based on the environment they are born into. From birth until their first day of school, these identities develop and change both internally and externally. Internally, changes occur with how the child sees themself and what their family instills in them. Externally, changes occur based on how people around them and the world sees them. When they begin some form of formal schooling, children often see their identities validated or invalidated and they may consciously or unconsciously start to alter their identities to fit in or become comfortable. Some children may even find themselves code-switching their identities when their dominant identities are not represented in the classroom. For example, if a child identifies as a girl and finds herself in a mathematics classroom where she feels boys are more valued—which is sadly still too often the case—she may feel the need to assimilate to the environment and at times engage in mathematics activities that may not be of interest to her; for example, she may choose to engage in continued back and forth with some of the boys to show that her method of solving a problem is accurate. Another example is when children enter a mathematics environment as the only person from their respective race/ethnicity. In this situation, they must decide whether they remain their authentic self or assimilate themselves to the environment they are entering. This may require them to leave part of their culture outside of the environment to obtain a level of acceptance. Therefore, children, especially those from historically excluded groups, must too often determine what they are willing to sacrifice to feel safe, respected, and fit in to the environments they are entering.

In addition to their identities, children also bring a rich set of resources to the mathematics classroom, better known as "funds of knowledge," which consists of the knowledge and skills they acquire from their families, culture, and everyday lived experiences. In their book *The Impact of Identity in K-8 Mathematics Learning and Teaching,* Julia Aguirre, Karen Mayfield-Ingram,

In addition to their identities, children also bring a rich set of resources to the mathematics classroom, better known as "funds of knowledge," which consists of the knowledge and skills they acquire from their families, culture, and everyday lived experiences.

and Danny Martin (2013) share five equity-based practices: (1) going deep with mathematics, (2) leveraging multiple mathematical competencies, (3) affirming mathematics learners' identities, (4) challenging spaces of marginality, and (5) drawing on multiple resources of knowledge. The last of these is what we're discussing here; these practices will be revisited in Chapter 5. Table 3.1 provides a brief list of characteristics you can use in the next Try This activity to help you get started with learning about your children in ways that include their funds of knowledge, their identities (which sometimes overlap), and their interests.

Table 3.1   *Examples of What Children Bring to the Classroom*

Funds of Knowledge	Identities	Interest
• Home language	• Race	• Hobbies
• Family values and traditions	• Ethnicity	• Things they like to do for fun
• Family outings	• Gender	• Things they are really good at
• Family occupations	• Language	• Books they like to read
• Educational activities	• Socioeconomic status	• Intended career path
• Favorite TV shows	• Family structure and a child's place within the family	• Thoughts about their community
• Entertainment	• Pieces of their culture a child values	• Thoughts about social justice
• Geography		• What world problem would they like to solve?
• Religion	•	
• Technology	•	•
•	•	•
•		•
•		•

Feel free to add identities and interests to the table and then consider using it as you complete Try This: My Children's Background to get to know the children in your classroom.

 TRY THIS: MY CHILDREN'S BACKGROUND

How well do you know the backgrounds of the children represented in your classroom?

Take a moment to identify the child in your classroom you know the best (Child A) and the child you know the least (Child B). Use the examples in Table 3.1 to catalog in the following chart what you know about each child.

	Funds of Knowledge	Identities	Interests
Child A			
Child B			

Compare and contrast the two children.

- How do you know this information is accurate? What evidence do you have?
- Would the children describe themselves in the same manner? How would their families describe them? How do you know?
- What is similar and what is different?

Reflect upon why you know one child more fully than the other.

- What will you do to learn more about the child you know the least about?
- What will you do to learn more about the child you know the most about?
- What will you do to learn more about the other children in your classroom?

 Available for download at https://qrs.ly/wbfixtr

Another way to learn about your children is to learn about their communities. Begin by examining your school's boundary map to determine what the neighborhoods are like that your children live in. Consider making home visits when possible. Visit local businesses, markets, community centers, and public spaces like libraries, parks, recreation centers, and so on, if and where they exist. Make note of how you see local children interacting in their neighborhood spaces. Use multiple sources to learn about the community in which the children reside, as every piece of knowledge will provide a different lens or perspective.

- Look at the demographic breakdown, specifically of race or ethnicity and socioeconomic status of the people who live within that boundary.

- Learn about the community through books, taking walks throughout the boundary area, talking with community elders, and watching local news.

By understanding the community, you begin to understand the environment and its richness of culture and opportunities. As you gain an understanding of the community, you begin to know the children under your purview and build relationships. You can also complete the Twenty Things to Know About Me activity in the Reflection and Action section at the end of this chapter to learn more about your children's backgrounds and their families. Now let's look at some other ways you can learn about your children's identities.

HOW DO I LEARN ABOUT CHILDREN THROUGH THEIR FAMILIES?

Now that you have given some thought to your children's background, take some time to verify the information you have gathered by having discussions with the children directly and talking to their families. Every opportunity you have to engage with families or caregivers should be taken advantage of. Here are some opportunities and ways to communicate with families:

➤ back-to-school events,

➤ home visits,

➤ calling families during the first or second week of school,

➤ open house events,

➤ extracurricular events,

➤ community events,

➤ newsletter updates with inquiries for families to respond.

The process of engaging families will not happen all at once, and these opportunities to talk will take place over time. It's also important to ensure it is a two-way street. Taking the time to authentically listen and gain insights *from* families as to who they are and what they desire for their children is as important as you communicating *to* them. Learning about children's identities is a continuous process and you should remember that some identities also evolve over time, thus allowing yourself and the children the opportunity to grow.

It's also important to ensure it is a two-way street. Taking the time to authentically listen and gain insights from families as to who they are and what they desire for their children is as important as you communicating to them.

How Do Various Identities Play a Particular Role in My Classroom?

Understanding various identities significantly impacts your TMSJ journey. It's crucial to comprehend how these identities influence your children's learning and school performance. Why does it matter so much? Deep knowledge and understanding of your children's diverse identities enables you to view all children inclusively and humanely. This perspective allows mathematics teachers to learn about, integrate, and lift up the differences, similarities, and intersections these children bring to the classroom. Through this understanding, educators can create robust, meaningful lessons that offer an affirming, social justice–oriented mathematics experience to all children.

WHAT ROLE DO RACE, ETHNICITY, AND CULTURAL IDENTITY PLAY IN MY CLASSROOM?

Although it is important to consider a variety of identities, we want to first address race, ethnicity, and culture. Two of the most prevalent forces of division and injustice in Western society are race and ethnicity, as persons in power have historically used them as proxies to maintain power and cause division among folx—thus our reason for the intentional focus. To establish equitable, socially just environments in mathematics education, it's imperative to address race, ethnicity, and culture intentionally to ensure every child's background is understood, represented respectfully, and woven into our lesson narratives. Table 3.2 provides ways racial, ethnic, and cultural identities can impact your classroom.

Table 3.2   *Race, Ethnicity, and Culture in the Mathematics Classroom*

Teacher-Child Relationships
• *Cultural Sensitivity*: Educators' understanding and acknowledgment of diverse cultural backgrounds can positively impact their relationships with children and create a more inclusive classroom.
• *Representation*: Educators from similar racial or ethnic backgrounds as their children can serve as role models and bridge cultural gaps, fostering better rapport and understanding.
Classroom Dynamics
• *Group Dynamics*: Children from different racial or ethnic backgrounds might have diverse learning styles or cultural norms that affect how they interact within group work or classroom discussions.
• *Language and Communication*: Children whose first language isn't the language of instruction might face challenges and need specific forms of support in learning mathematical concepts or expressing their ideas.
Student Engagement
• *Stereotypes and Bias*: Educators' expectations of children's mathematical abilities can be influenced by preconceived notions or biases they have about certain racial or ethnic groups.
• *Identity Threat*: Stereotypes suggesting certain racial or ethnic groups are not strong in math may cause children from those groups to experience anxiety.
• *Cultural Relevance*: Relating mathematical content to children's cultural experiences can influence their engagement and interest in the subject.

As you consider how race, ethnicity, and culture play a role in your classroom, use the following to take inventory of your classroom.

▶ What images are portrayed through pictures on the walls, artifacts on shelves, or through handouts that children receive?

- ➤ What sounds, if any, do children hear when they enter your classroom? Is music playing? If yes, what genre of music?

- ➤ If objects are in your room (e.g., toys, touchable artifacts), what and/or who do they represent?

It all comes down to a teacher's ability to clearly "see" the children sitting in their classrooms for everything about who they are as individuals and everything about what ethnic and cultural values, experiences, and influences have so far formed who they are.

"I DON'T SEE COLOR"

Often, people operating in white-centered educational environments do not *see* melanated children, and they historically excluded those children from mathematics, figuratively and literally—even though their ancestors developed the foundational components of the mathematics concepts we use today (Okun, 1999). When we say they are not "seen," we are not referring to the physical sense of seeing but rather to whether they are value, affirmed, and authentically engaged. Too often in mathematics education environments, melanated children and historically excluded children are afterthoughts and not centered in the experiences.

In the United States, thanks to a historic cultural lens based in white supremacy, most people—white or otherwise—are taught they should not "see" the race or ethnicity of the nonwhite children, even though melanated children increasingly populate their classrooms. That is, they are guided to ignore, suppress, or negate—as much as possible—the cultural attributes associated with the racial and ethnic identities of their nonwhite children and to focus instead on seeing those children as individual "persons," lacking any culturally differentiating characteristics from their white peers. In effect, society conditions teachers (particularly white teachers) to *deracialize* and *de-ethnicize* their nonwhite children and to view them, to some extent and for certain purposes, as nonwhite versions of white children (Choi, 2008). The intention behind this colorblind approach is to minimize or avoid inserting their overt or covert biases into their instruction and instead attempt to treat all children equally. Equal, by definition, means to treat children in the same manner as all others. However, student achievement data shows this to be a flawed approach for several reasons. First, it doesn't actually help. There are still significant student achievement gaps among children (National Center for Education Statistics, 2022). No matter how much teachers, schools, and systems may think they are treating students equally, data about access to rich mathematics and student success in mathematics tell us that racial gaps continue to be perpetuated.

Second, this colorblindness is harmful. Not only does it often lead teachers to misjudge, underappreciate, and underestimate the unique and important knowledge and experiences melanated nonwhite children bring to the classroom, but colorblindness also causes teachers to fail to effectively incorporate their melanated nonwhite children's identities into classroom instruction. It results in a form of erasure rather than an opportunity to build on what those children can bring into the classroom and what they need to propel their learning. The phrase "I don't see color" actually means "I don't see some of my children for their whole selves." At its worst, this approach also often leads those children to ask: "Do I even belong here?"

Summary

This chapter is one of the most important in the book as it focuses your attention on your children. The work to learn about your children—their backgrounds, funds of knowledge, identities they bring into the classroom, interests, and strengths—is not an easy lift but it's one that must be taken, especially if you are committed to creating a classroom environment and mathematical experiences in which your children can develop a positive mathematical identity and sense of agency. We encourage you to learn a little something about each of your children and select a few to really get to know by using the tools in this chapter to take a deep dive. Trust us; the time and energy are well worth it.

To hear more from the authors about their reflections on understanding your children, listen to this **conversation with Dr. Childs and Dr. Staley.**

qrs.ly/qkffulp

REFLECT

Our children and the relationships formed—children and teacher, children and children—are necessary for TMSJ. Consider the following questions as you think about your children and steps you can take to connect with and learn more about them.

- What steps can you take to learn more about your children's funds of knowledge, strengths, and interests?

- How do race, ethnicity, culture, and other identities play roles in your classroom?

_____

_____

_____

_____

_____

_____

 **ACT**

1. Add your key takeaways and next steps to your TMSJ Action Plan.

2. Select one to three children and learn about their backgrounds and identify their funds of knowledge and strengths. Record the information in the chart below. Be intentional in using this information as you foster a nurturing TMSJ setting that is inviting and imbued with a sense of belonging for your children. (See also Tables 3.1 and 3.2.)

3. Twenty Things to Know About Me: Have each child list ten things they want you to know about them. Have the child's family member or guardian list ten things they want you to know about the child.

Use this list to build connections with each child. Cultivating these relationships will require time and effort. Regardless of whether you teach at the elementary, middle school, or high school level, fostering relationships is imperative in a TMSJ classroom. For some children, rapport may form instantly, while for others, it might take days, weeks, or even months. There's also a possibility that a relationship may not develop, and that's acceptable. By utilizing the Twenty Things to Know About Me, you'll uncover their strengths and the valuable knowledge they bring into the classroom. This insight will be vital as you construct meaningful and impactful TMSJ experiences.

My Children's Strengths

	Child A	Child B	Child C
Background			
Funds of knowledge			
Dispositions			
Mathematical processes and practices they are strong in			
Mathematical content they are strong in			
One way I plan to use this information			

online resources Available for download at https://qrs.ly/wbfixtr

Where to Next?

Before you continue on your journey, we invite you to pause and reflect on Part 1: A Social Justice Mathematics Teaching Framework. These three chapters have laid the foundation for the work that is needed to transform the teaching and learning of mathematics in your classroom and center teaching mathematics for social justice.

▶ Chapter 1 provided an overview of the social justice mathematics continuum, which connects classroom equity and culturally relevant pedagogy, culturally responsive teaching, and teaching mathematics for social justice.

▶ Chapter 2 asks us to look in the mirror and examine how our own identities shape us as mathematics educators. It's not enough to be aware that we all have biases; we must acknowledge how they might show up in our classrooms and, especially, in interactions with our children.

▶ Chapter 3 gets to the heart of why we wrote this book, our children. This chapter challenges you to truly get to know the children in your classroom.

PART 2

THE MATHEMATICS EXPERIENCE

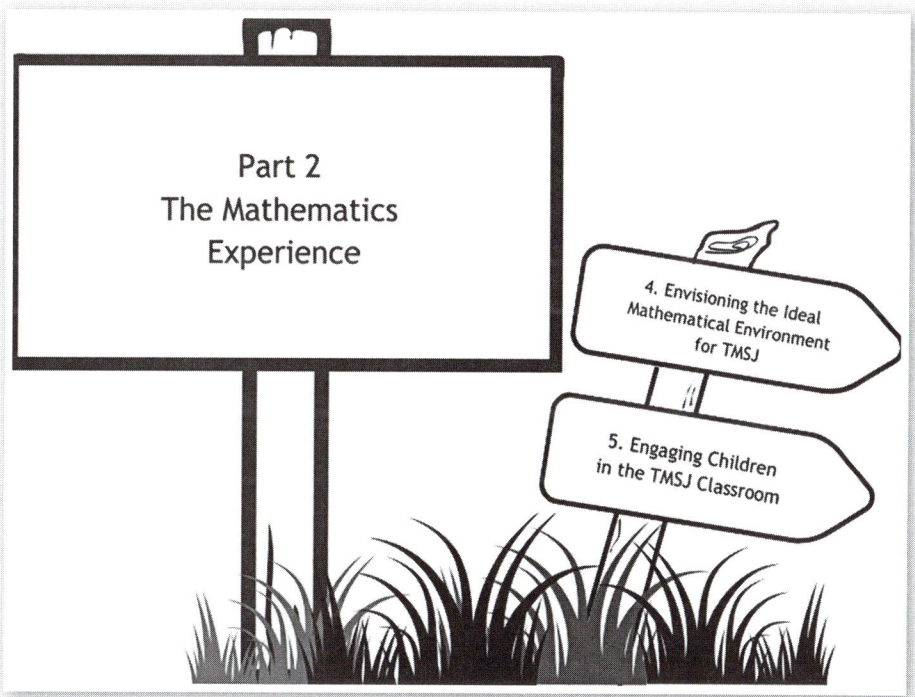

Part 2
The Mathematics
Experience

4. Envisioning the Ideal Mathematical Environment for TMSJ

5. Engaging Children in the TMSJ Classroom

In Part 1, the focus was on providing you with the foundational tools needed to begin your TMSJ journey, with a keen focus on reflecting upon who you are and how you enter this work. In Part 2, we build upon this work by laying the foundation for impacting the classroom environment. In Chapter 4, we will envision the ideal environment for TMSJ and in Chapter 5, make sense of effectively engaging children in the TMSJ classroom.

ENVISIONING THE IDEAL MATHEMATICAL ENVIRONMENT FOR TMSJ

In the previous two chapters, we asked you to look closely at yourself and learn about the children in your classroom. The work of TMSJ starts with oneself. Now, it's time to consider the ideal mathematical environment for TMSJ. This will require you to reflect on your classroom and its culture to determine actions you will take to create a space that encourages equal participation and where all your children are safe socially, emotionally, and academically. Figure 4.1 shows the cornerstones that form the foundation for TMSJ, grounded in an equitable and engaging mathematical environment:

▶ cultural responsiveness

▶ community

▶ collaboration

▶ engagement

In Chapter 4, we focus on the first three cornerstones and will continue with engagement in Chapter 5.

Figure 4.1   *Cornerstones of the TMSJ Mathematical Environment*

Each of us engaged in mathematics experiences while growing up. Some of us loved the experience while others disliked it. However, the experience shaped each of us as mathematics educators. Many of us experienced mathematics mostly through the assimilationist environment. An assimilationist environment is one that requires children to learn mathematics through the Eurocentric lens of the dominant North American culture—one that centers white-identifying people and the white experience and asks children to *assimilate* to that perspective. In this environment, instruction focuses primarily on ancient Greek and European mathematical precepts, generally references white mathematicians as the originators and arbiters of mathematics, and relies heavily on direct instruction, rote memorization, and mimicking. In this environment, our teacher would demonstrate how to solve a mathematical problem, which we would then replicate, and then we practiced independently—also known as "I Do, We Do, You Do." For much of the last three decades, this was a common—if not predominant—pedagogical approach in schools. It had less of a focus on contextualized or situational problem-solving designed to build and demonstrate thinking and more of a focus on procedural tasks that children were expected to replicate over and over for practice and then replicate again on a test to demonstrate proficiency.

One drawback of teaching mathematics through an assimilationist lens is that it robs children of seeing themselves and their loved ones in the mathematics. It neutralizes the subject into a stark and limited narrative of how numbers work together—and who figured that out—but it does little to help children understand how mathematics relates to and is applicable to their lives. This is one of the goals of TMSJ.

Before discussing the ideal TMSJ mathematical environment, we invite you to reflect on the kind of classroom experience you had as a child or know of from your professional experience.

 TRY THIS: REFLECTING ON YOUR EXPERIENCE

Take a moment to reflect on how mathematics learning is structured in your classroom, your building, or your district. Is this structure similar to or different from your experience as a student? Explain.

As you compare and contrast your school experience with your current environment, consider whether your current environment is what you imagined it would be. What do you love about it? What would you like to change?

Now, let's look at the first of the four cornerstones of the ideal environment for TMSJ: cultural responsiveness.

Cultural Responsiveness

As defined in Chapter 1, culturally responsive teaching focuses on centering children's cultural identity in the mathematics experience (Gay, 2002). Using this definition as a foundation, we are positing a concept of culturally responsive mathematics environments in which mathematics instruction is shaped around the unique thinking, knowledge, and problem-solving skills of diverse groups of children. In addition, as described in Chapter 1, this is a constructivist approach in which teachers encourage learning through engagement and exploration, and it reflects the diverse cultural backgrounds children bring to the classroom. Chapter 3 stressed the importance of learning about your children, which is a necessary step if you are to create a culturally responsive environment. Table 4.1 highlights select elements of cultural responsiveness, based on Gay's (2002) culturally relevant instruction, that should be prominent for TMSJ. We also describe what it looks like in the mathematics classroom.

Table 4.1 *Elements of Cultural Responsiveness*

Element	What It Looks Like in the Mathematics Classroom
Positive perspectives on families	Educators view families through an asset-based lens, always believing that families are giving their best effort.
Learning within the context of culture	When engaging children in mathematics tasks, educators seek ways to authentically integrate components of children's respective cultures.
Student-centered instruction	Educators consistently seek ways to focus on children throughout the lesson, acting as a facilitator of building knowledge rather than a bearer of all knowledge to be delivered.
Culturally mediated instruction	Educators incorporate diverse ways of knowing, understanding, and representing information.
Reshaping the curriculum	Educators consider ways to decenter whiteness in the curriculum experience, ways to incorporate the lived experiences of all children, and ways to allow them to showcase their thinking.

Source: Adapted from Gay (2018).

As you consider which elements need to be elevated in your classroom, we remind you to focus your attention on your children so that you can create an environment that is not just a space for learning mathematics but also a safe and supportive community.

Community

The second cornerstone, community, involves creating a space in which all children can "make sense of their and others' lived experiences and understand their and others' agency" (Conway et al., 2022). Community begins with using an understanding of your children's identities to shape classroom values and commitments. This understanding also guides your selection of instructional practices that are student centered and student focused, to foster a sense of belonging in which all children are seen, heard, cared for, and valued as creative thinkers and doers of mathematics. There are two components to building community: values and commitments.

> Community begins with using an understanding of your children's identities to shape classroom values and commitments.

BEGIN WITH CLASSROOM VALUES

As you establish classroom values, consider focusing on those that have the potential to deepen your children's understanding of the social justice domains (identity, diversity, justice, and action) as well as develop their social, emotional, and cognitive skills and competencies. In Table 4.2, we integrate several social justice outcomes (Learning for Justice, 2016) with selected competencies from the Collaborative for Academic, Social and Emotional Learning (CASEL) Framework.

Table 4.2 *How Social-Emotional Competencies and Social Justice Outcomes Shape Mathematics Classroom Values*

Social-Emotional Competency	Social Justice Outcomes in the Context of Mathematics Teaching and Learning
Self-awareness: The ability to understand one's own emotions, thoughts, and values and how they influence behavior in different contexts.	*I-1. Students will develop positive social identities based on their membership in multiple groups in society.* **Teachers** recenter identities, perspectives, and knowledge traditions that have often been silenced. **Children** recognize that people's multiple identities interact and create unique and complex individuals that contribute to their learning of mathematics. *D-6. Students will express comfort with people who are both similar to and different from them and engage respectfully with all people.* **Teachers** design and implement a curriculum that honors diversity in mathematical reasoning, sensemaking, and multiple forms of engagement to promote individual and collective learning. **Children** express comfort in working with and learning from people who are both similar to and different from them and engage respectfully in collaborative work and discussion.

(Continued)

Social-Emotional Competency	Social Justice Outcomes in the Context of Mathematics Teaching and Learning
Social awareness: The ability to understand the perspectives of and empathize with others, including those from diverse backgrounds, cultures, and contexts.	*I-2. Students will develop language and historical and cultural knowledge that affirm and accurately describe their membership in multiple identity groups.* **Teachers** attend to and honor students' multiple social identities in curricular design and its implementation. **Children** develop language and historical and cultural knowledge to affirm and describe their membership in multiple identity groups and their contributions to mathematics. *D-8. Students will respectfully express curiosity about the history and lived experiences of others and will exchange ideas and beliefs in an open-minded way.* **Teachers** create multidimensional classrooms, raising students' expectations for contributions from each and every child. **Children** express curiosity about the mathematical contributions and experiences of others and exchange ideas and perspectives in an open-minded way.
Relationship skills: The ability to establish and maintain healthy and supportive relationships and to effectively navigate settings with diverse individuals and groups.	*I-4. Students will express pride, confidence, and healthy self-esteem without denying the value and dignity of other people.* **Teachers** view students as competent mathematical beings whose lived experiences and community and cultural ways of knowing are leveraged during mathematics instruction. **Children** express self-love, pride, confidence, and healthy self-esteem about themselves and their community as mathematical thinkers and learners. *D-9. Students will respond to diversity by building empathy, respect, understanding, and connection.* **Teachers** deconstruct stereotypes about students' mathematical identities and who can and cannot do mathematics. **Children** respond to diversity by building respect, understanding, connections, and empathy for different ways of knowing and being in mathematics classrooms.

Sources: SEL directly from CASEL (n.d.); SJ Outcomes from Teaching Tolerance (2016); Context from Bartell, et al. (2022) and Koestler et al. (2022).

For additional guidance on social-emotional and academic development, visit Stride 3: Creating Conditions to Thrive (https://equitablemath.org) and Integrating Social, Emotional and Academic Development (SEAD): An Action Guide for School Leadership Teams (https://www.aspeninstitute.org/publications/integrating-social-emotional-and-academic-development-sead-an-action-guide-for-school-leadership-teams).

CLASSROOM COMMITMENTS BRING VALUES TO LIFE

Classroom values set the stage for you to now work with your children to develop classroom commitments. More than rules for classroom management, classroom commitments require buy-in from students and are the norms that help establish an environment where children can be themselves and learn from their classmates. Commitments should

➤ evolve from the classroom values and emphasize interactions between all individuals (teacher-child, child-child, child-other educators/adults) so that everyone can feel safe to actively participate in the mathematics classroom;

- establish norms that foster a sense of belonging, welcome multiple perspectives, and encourages diverse views;

- be co-created, reviewed, and revised in an appropriate manner based on your children's ages.

Table 4.3 provides considerations for these three commitment areas and ways they may show up in the mathematics classroom.

Table 4.3 *Social Justice at the Center of Mathematics Classroom Community Commitments*

Commitments Should	Social Justice Standards in the Context of Mathematics Teaching and Learning
Emphasize Interactions • Model respectful ways of listening, questioning, and valuing each other's mathematical ideas. • Set expectations for sharing solutions and asking questions. • Be mindful of the ways all individuals respond to each other. • Provide guidelines for listening and question stems for children to use when critiquing the reasoning of others.	*J-11. Students will recognize stereotypes and relate to people as individuals rather than representatives of groups.* **Teachers** locate causes of inequalities in social conditions (e.g., tracking, ability grouping, Eurocentric curriculum) rather than believe conditions are inherent within individuals. Teachers provide students with opportunities to use mathematics to explore these causes. **Students** recognize stereotypes and pervasive myths around what mathematics is and what it means to know and be good at mathematics.
Foster Belonging • Ensure accurate pronunciation of children's names and use of preferred pronouns. • Have a list of expectations for learning in mathematics class created by children. • Ensure mathematical discourse begins with acknowledging each voice. • Monitor adult-child and child-child interactions and address microaggressions. • Have children generate a list of expectations that can help them be successful.	*J-14. Students will recognize that power and privilege influence relationships on interpersonal, intergroup, and institutional levels and consider how they have been affected by those dynamics.* **Teachers** explicitly shift the power dynamic between children-teacher and children-children by centering identities, perspectives, and knowledge traditions that have often been silenced. **Students** recognize that power and privilege influence relationships on interpersonal, intergroup, and institutional levels and consider how they have been affected by those dynamics in their mathematics learning experiences and in the world.
Be Co-Created • Present a set of norms and gather feedback to adjust. • Provide processing time and revisit norms.	*A-17. Students will recognize their own responsibility to stand up to exclusion, prejudice, and injustice.* **Teachers** provide students with a consistent opportunity to recognize their own responsibility to stand up to exclusion, prejudice, and injustice. **Students** plan and carry out collective action using mathematics as a tool to address injustice in the world.

Sources: Teaching for Tolerance (2016); Bartell et al. (2022); Koestler et al. (2022).

Collaboration

Collaboration is one of the learning and innovation skills in the Framework for 21st Century Learning, which also includes critical thinking, communication, and creativity.

As you work to integrate cultural responsiveness into your classroom to create a sense of community, you must intentionally construct collaborative opportunities for your children. For the purpose of this discussion, we define collaboration as "a process through which learners at various performance levels work together in small groups toward a common goal" (UNESCO-UNEVOC International Centre, n.d.).

Collaboration is one of the learning and innovation skills in the *Framework for 21st Century Learning*, which also includes critical thinking, communication, and creativity (P21, 2019). This third cornerstone is essential in a TMSJ classroom, as it supports children's learning as they work in groups to organize their efforts to reach a goal—that of learning and using mathematics to solve problems, especially real-life problems that come from their school or community. In a learner-centered approach, collaborative learning fosters positive interdependence, individual accountability, and interpersonal skills among children (Key Differences, 2024). Table 4.4 lists the Center for Teaching Innovation (2024) benefits of collaborative learning, and we have added a few ways this shows up in the mathematics classroom.

Table 4.4 *Benefits of Collaborative Learning*

What Collaborative Learning Leads to	Ways This Shows Up for Children in the Mathematics Classroom
Development of higher-level thinking, oral communication, self-management, and leadership skills	Children share mathematical thinking and solution strategies. They critique the reasoning of others and receive/provide feedback.
Promotion of children-teacher interaction	Children bring their existing funds of knowledge (culture, contexts, language, and experiences) as assets to gain understanding of context.
Increase in child retention, self-esteem, and responsibility	They develop positive mathematics and social identities. Mathematical agency is enhanced through regular collaborative opportunities to work on tasks that require cross-curricular connections, reasoning, and sense making.
Exposure to and an increase in understanding of diverse perspectives	Children learn about their peers' culture and lived experiences. They engage respectfully with peers who are similar to and different from them.
Preparation for real-life social and employment situations	Children prepare for mathematics beyond the classroom by engaging in learning experiences in which application of knowledge is needed in conjunction with social skills.

Source: Center for Teaching Innovation (2024).

Take a moment to reflect on the level of collaboration in your classroom environment.

Check In

Collaboration in My Classroom

Describe what collaboration looks and sounds like in your current classroom. What are the children doing? What are you doing?

HELPING CHILDREN COLLABORATE

It may seem intuitive that children can naturally collaborate. For example, we see them on the playground, in the courtyard, in the lunchroom, and in the community engaging with each other in a multitude of ways. Often, children are given academic tasks and are told to work with a peer. However, through an academic lens, the collaboration skills used outside of the mathematics classroom do not always transfer into the classroom.

We must help nurture them in this process, giving them the tools needed to know how to collaborate productively. Consider the children under your purview and identify the ways they best collaborate. For example,

> Game play: When children are playing games, they become deeply involved in achieving the goal of the game. Often they provide each other strategies and tips to be successful in the game; in addition, they share previously experienced scenarios. When game playing, children seek shortcuts and ways to maximize opportunities and extend game play.

> Storytelling: When children tell stories, especially in groups, notice their body language and their peers' engagement as they are listening. Children will, at times, interject with ideas or hypotheses for the stories. At times they will intentionally or unintentionally change the plot of the story. However, the children are keenly invested in the narrating of the story and listening to the story and celebrating the end of the story.

How can they improve in their collaboration? To answer this question, first look at the level of interaction in each of these situations. In most cases, one might observe a change in the volume of voices (increase or decrease), body language and gestures, facial expressions, and proximity of individuals (close or spread apart). Now consider the connection among the children involved in the interaction, which is often much harder to see. Each child's level of collaboration may vary based on their relationship with those involved. These observations from outside of the classroom can help you lean into the ways children collaborate best when introducing new mathematical topics. To further help children collaborate, one must intentionally teach children collaborative skills for working in groups. Table 4.5 provides some key features to establish a culture of collaboration.

Table 4.5   *Key Features to Create a Collaborative Culture*

Key Features	Teacher Actions	Ways This Shows Up for Children in the Mathematics Classroom
Setting goals	Require children to set interaction goals to guide the collaboration.	Children list individual goals. During collective time at the beginning of the collaboration, collect goals from the children and record them on the board for the classroom community to seek to achieve.
Varying group structures	Ensure collaborative groups are designed with educational and social goals in mind to optimize the learning experience.	Create fluid groups based upon discussion topic(s). Allow children from different cultural backgrounds opportunities to collaborate.
Listening	Provide children time to make sense of others' strategies and perspectives by using their auditory skills to process communication.	Provide children individual time to explain to you their thinking. During whole class discussions, ensure all children are attentive when their classmates are speaking.
Asking questions (see Chapter 5 for additional details)	Encourage children to ask questions and understand that no one person is the sole knowledge bearer. Everyone has something to contribute, and asking questions helps the teacher identify and support children's understanding.	Provide children watch time to note at least one question they have. Ensure all children throughout the experience are provided an opportunity to ask a question of the whole group.
Encouraging innovation	Encourage innovative thinking and brainstorming of multiple ways to approach and solve a task. Empower children to share their thinking.	Encourage children to brainstorm multiple problem-solving strategies. Provide opportunities for children to collaborate and seek efficient solution strategies.
Recognizing or rewarding	Recognize or reward the actual collaboration as much as the task completion.	Verbally provide children independent praise. In whole-group settings, acknowledge group collaboration efforts at all levels.

As you consider the key features, we want to look more closely at the topic of varying group structures, as this is an area where there are vast differences in structure in classrooms we see regularly. We turn your attention to research from Peter Liljedahl, author of *Building Thinking Classrooms in Mathematics, Grades K-12: 14 Practices to Enhance Learning* (2020), on group structures, to increase the educational benefits of collaborative groups.

VISIBLY RANDOM GROUPING TO FOSTER COLLABORATION

Collaboration—or what commonly shows up as collaborative groups—has educational (pedagogy, productivity, and peacefulness) and social (diversity, integration, and socialization) purposes (Liljedahl, 2020). Collaborative educational environments do not happen through happenstance; they occur because the classroom teacher understands the value of collaboration and is intentional about helping children develop their collaborative capacity. Liljedahl observed that there is often a conflict between the teacher's goals when grouping children and children's goals when participating in groups, especially when groups are self-selected. He encourages the use of visibly random grouping to help increase children's engagement and thinking. Table 4.6 provides an overview of this grouping method for your consideration.

Table 4.6   *Visibly Random Grouping*

Process	Frequent regrouping (daily or even hourly) using a random method that is visible to children (i.e., having children draw a card from a deck, pulling names from a bag)
Rationale	Groups are generative and have redundancy and diversity. • Redundancy: things that groups of children have in common—language, interests, experiences, knowledge • Diversity: things that children bring to the group that are not shared—different ideas, viewpoints, perspectives, representations Visibility of the randomness is crucial because if students believe groups are not randomly selected, they automatically assume their "role" in the group (leader vs. follower) based on their impressions of how the teacher views their ability.
Optimal group size to maximize the balance of redundancy and diversity	Grades K-2: 2 children Grades 3-12: 3 children
Benefits	Increased willingness to collaborate Elimination of social barriers Increased knowledge mobility Increased enthusiasm for mathematics learning Reduced social stress Development of empathy

Source: Liljedahl (2020).

Now take a look back at your response for the collaboration Check In: Collaboration in My Classroom and consider the following for collaborative activities you described:

➤ What were the goals for the collaborative activities?

➤ Which activities had an educational or social goal?

➤ What adjustments can you make to enhance the learning experiences for your children?

PREPARING CHILDREN FOR GOING BEYOND THE CLASSROOM

In the world outside of formal schooling, most relationships and even most careers require people to collaborate in some form on everything from projects at work, to community organizing, to planning events, and so much more. If school is preparing children for this world, activities should be integrated that provide them the opportunity to develop their collaborative capacity and skills. These opportunities should vary from engaging in discourse daily to engaging in projects in which each group member has a vital role to play that requires cooperation from the start to the finish, or some combination of these elements. Collaborative activities should be a natural component of children's educational experience. Take a moment to complete the Check In activity.

Check In

Make a list of activities that require collaboration and those that can be done independently. Use your imagination to consider career, work, and nonwork occasions and activities.

Collaboration or Independent Activity?

Activities That Require Collaboration	Activities That Can Be Done Independently

online resources 🔗 Available for download at https://qrs.ly/wbfixtr

Many of your current classroom mathematics activities provide opportunities for children to collaborate and engage in discourse. However, the question we must continually ask ourselves is "Are we maximizing collaboration and engagement opportunities?" This can range from a simple turn and talk during a lesson to an opportunity for children to create a collaborative project. Consider using Check In: Collaboration or Independent Activity? with your children and allow time for them to research some careers and other activities to show why collaboration is an essential skill they should develop while in school. The Mathematical Association of America's Math Career Resource Center (https://mathcareers.maa.org/) is a good place to start. As you create collaborative opportunities and activities for your children, remember that the key is not recreating new things from scratch but rather focusing on opportunities for children to further develop their collaboration skills.

Summary

As we shared in Chapter 1, TMSJ is more than just integrating social justice issues into your mathematics lessons. The mathematical environment is a vital component of TMSJ. In this chapter, we focused on three of the four cornerstones that form the foundation for an equitable and engaging mathematical environment: culturally responsive learning environment, community, and collaboration. Each of these speaks to the importance of creating a safe space with and for your children where they have an opportunity to thrive socially, emotionally, and academically in the mathematics classroom while developing a positive mathematical identity and sense of agency.

To hear more from the authors about their reflections envisioning the ideal mathematics environment, listen to this **conversation with Dr. Childs and Dr. Staley.**

qrs.ly/zjffulr

REFLECT

Reflect on the three cornerstones—culturally responsive learning environment, community, and collaboration—to determine areas where adjustments are needed in your classroom. Consider the following questions as you view each cornerstone from the eyes of your children in your current mathematics classroom:

- How might an increased focus on cultural responsiveness improve the sense of community and level of collaboration in your classroom?

- What changes do you need to make to improve the sense of community?

c Is your current environment conducive to collaboration? Why or why not? How can you make it a more collaborative environment?

_____

_____

_____

_____

_____

_____

_____

_____

_____

📢 ACT

1. Add your key takeaways and next steps to your TMSJ Action Plan.

2. Now take a moment to rate your classroom mathematics environment (1 = Beginner; 2 = Intermediate; 3 = Advanced). Refer to the chapter tables as needed.

_____

_____

_____

_____

_____

_____

_____

_____

_____

My Classroom Environment

Key Features	Rating	Notes
Enhanced communication		
Listening		
Setting goals		
Leveraging strengths		
Asking questions		
Encouraging innovation		
Recognizing or rewarding		

online resources 🔎 Available for download at https://qrs.ly/wbfixtr

Where to Next?

In Chapter 5, we will address the fourth cornerstone, student engagement. Student engagement gets to the core of your instructional design: *what* mathematical tasks and activities you select, *how* you design the instructional approaches and routines, and *who* your children are as they showcase their thinking.

5 ENGAGING CHILDREN IN THE TMSJ CLASSROOM

In the previous chapter, we focused on the first three cornerstones of a TMSJ environment: cultural responsiveness, community, and collaboration. We now focus on the fourth cornerstone: the importance of engaging our children in the TMSJ classroom. This work begins with a commitment to

- equitable mathematics teaching;

- a balanced instructional approach that includes building procedural fluency, conceptual understanding, and problem-solving;

- utilizing rich problem-solving tasks; and

- showcasing children's thinking.

Equitable Mathematics Teaching

The foundation for equitable teaching is rooted in research-informed mathematics teaching practices that consider the behavior and actions of teachers and children. We offer that the NCTM Mathematics Teaching Practices and Standards for Mathematical Practice (or similar process and practice standards), when integrated with the five equity-based mathematics teaching practices (Aguirre et al., 2013), form a solid foundation for the equitable teaching and learning of mathematics (see Table 5.1).

Table 5.1 *Foundation for Equitable Teaching*

Equity-Based Mathematics Teaching Practices (Aguirre et al., 2013)	Mathematics Teaching Practices (NCTM, 2014)	Standards for Mathematical Practice (SMP) (NGACBP & CCSSO, 2010)
• Go deep with mathematics. • Leverage multiple mathematical competencies. • Affirm mathematics learners' identities. • Challenge spaces of marginality. • Draw on multiple resources of knowledge (math, culture, language, family, community).	• Establish mathematical goals to focus learning. • Implement tasks that promote reasoning and problem-solving. • Use and connect mathematics representations. • Facilitate meaningful mathematics discourse. • Pose purposeful questions. • Build procedural fluency from conceptual understanding. • Support productive struggle in mathematics. • Elicit and use evidence of student thinking.	• Make sense of problems and persevere in solving them. • Reason abstractly and quantitatively. • Construct viable arguments and critique the reasoning of others. • Model with mathematics. • Use appropriate tools strategically. • Attend to precision. • Look for and make use of structure. • Look for and express regularity in repeated reasoning.

Equitable mathematics teaching is the foundation for creating a mathematics classroom with a balanced pedagogical approach that utilizes rich problem-solving tasks and showcases student thinking.

A Balanced Approach to Nurturing Children's Understanding

Children are just like adults in that they have a genuine desire to engage in activities that pique their interests. As discussed in Chapter 3, children enter our classrooms with a multitude of interests and experiences, but during mathematics lessons they often must suppress these interests as they grapple with unfamiliar contexts and abstract concepts that are hard to make sense of when disconnected from their interests and experiences. Often, teachers unwittingly attempt to intertwine contexts based on their own interests and project onto children their excitement for the interest. In addition, many teachers focus on ensuring the daily content is aligned to the standardized assessment, and they often unconsciously neglect the role of context to support children with gaining an understanding of the mathematics concept for the lesson. When this happens, the purpose of the learning becomes about passing an assessment rather than allowing children the opportunity to fully make sense of a concept through the lens of their own interests. It is imperative that both context and content receive equal intensity as it relates to creating engaging experiences. Figure 5.1 describes this balance in more depth.

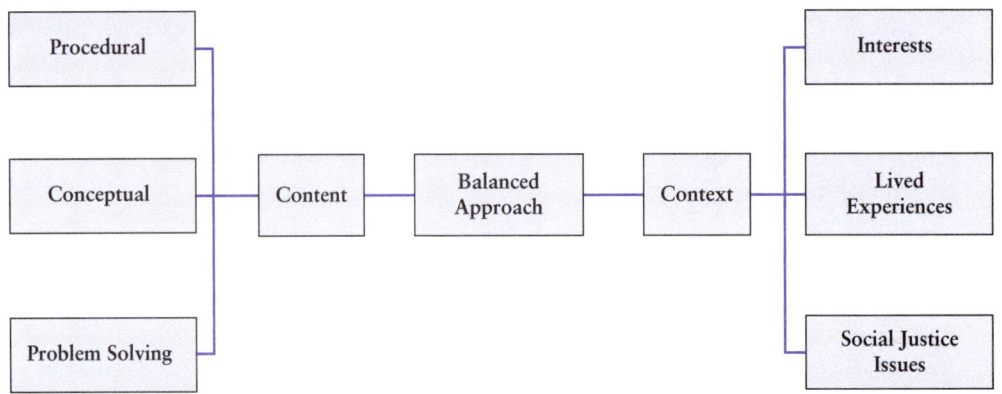

Figure 5.1 *A Balanced Approach for Engaging Students*

BALANCING CONTEXT WITH CONTENT

When thinking about the problems or tasks your children are challenged to solve, ask "Are the contexts contrived by a textbook publisher, based on teacher interests, or based on children's interests? Are you authentically including children's interests or making assumptions?" A context-first approach focuses on the scenarios that will be used to engage children in the content by tapping into their actual interest. This approach de-emphasizes the role of content and values the children's voice when determining context for the lessons. We suggest that you use information gathered about your children (see Chapter 3 for My Children's Background) to identify topics that connect to your children's interests, lived experiences, and school or community issues. When considering a social justice context, it is critical to focus on children's interests and not assumed interests. In mathematics class, teachers may be tempted to use social justice contexts they deem exciting, important, or both. These topics are often areas the teacher can connect to their prior experience or interest. When this occurs, it often requires children to figure out a way to make sense of the issue, especially if they have no prior knowledge of or interest in the topic. Take a moment to complete the Check In to identify examples of context you have used in your classroom.

 **Check In**

Non-Textbook Examples

List the last three non-textbook examples you have used in your classroom.

- What was the context?
- Did the children or you as teacher choose the context?
- How engaged were the children in the mathematical task? In the context?
- Going forward, what can you do differently?

Using children's interests and lived experiences for context will require you to be intentional and strategic about when to do this as you consider your curriculum materials. Some materials may have problems and tasks with context that makes it easier for you to make the connections and, at times, you may need to adjust a curricular task context for the lesson. For example, contexts that are relevant to the children under your purview can easily be incorporated, while others will need to be modified to provide children an entry point of understanding into the task. Almost any problem can be recontextualized into something children are interested in. Use the following questions as you consider context to use in the lesson:

- Is there a contextual example (connected to my children) in my curriculum materials that can be used?

- If using a non-textbook example, is this context children or teacher generated?

It is important to maintain the level of rigor of the content when incorporating context from your children, especially when a social justice context is used (see Chapter 6 for more details). In the next section, we turn our attention to the mathematical content.

BALANCING INSTRUCTIONAL APPROACHES FOR CONTENT

In a content-first approach, one focuses on "the mathematics" as it relates to the numbers, operations being performed, and the intrinsic value of the material being covered. The content-first approach assumes the child will find joy in performing the mathematics itself and finding a solution to the given task. According to Boaler (1993), this de-emphasis on the context aligns with a mathematics educator traditionalist mindset that regardless of the context "the mathematics" content takes precedence, which is contrary to a TMSJ educator approach.

Trends in teaching and learning mathematics have historically shifted back and forth from a procedural to a conceptual focus, shaped by accompanying classroom pedagogies that have also shifted. Today's classrooms tend to fall more heavily into one of three pedagogical approaches based on the beliefs of the classroom teacher, the teacher's experiences as a learner, selected curriculum resources, or the dictates of the district or school in which they work. Understand that each approach has value, but to create the ideal TMSJ environment teachers must establish a balanced approach that incorporates all three. Take a moment to explore the three pedagogical approaches—procedure-driven, concept-focused, and problem-solving—and how they come together to create a balanced instructional approach for content that is grounded in the grade-level standards and designed to support your children's needs (see Table 5.2).

Table 5.2 *Pedagogical Approaches for TMSJ*

Pedagogical Approach	Teacher Look-Fors	Children Look-Fors
Procedure-Driven Classrooms A procedure-driven classroom focuses on children learning the appropriate procedures and applying them to mathematics tasks.	Uses a transmission method: first, telling and demonstrating to children what needs to be done by sharing the steps or procedures to complete a problem. Focuses on the correct solution, not the process of finding the solution. Children are often taught "tricks," mnemonic devices, and other strategies to quickly find answers to tasks.	Children have a collaborative opportunity to engage in a mathematics problem-solving process.
Concept-Focused Classrooms A concept-focused classroom centers on children gaining an understanding of the relationships or foundational ideas of a mathematics concept.	Acts as the facilitator, allowing children space and opportunity to explore the beauty of mathematics, look for patterns, use multiple representations (e.g., concrete materials, illustrations or diagrams, symbols, words, and actions) to work through problems that have multiple possible solution pathways, and form generalizations about the mathematics. Asks thought-provoking questions to challenge and encourage children's thinking and helps them consolidate what they have learned once they've had an opportunity to explore together.	Use sense-making skills and strategies to approach a given task.
Problem-Solving Classrooms Problem-solving classrooms unlock mathematics through an interconnection of children's knowledge and understanding of mathematical ideas.	Classroom environment is rooted in children's engagement in rich problem-solving tasks. Tasks are open-ended and provide children the space and opportunity to showcase their thinking. Encourages diversity of thought and new ways of thinking and approaching the task.	Collaborate with others and share their thinking.

A balanced instructional approach—pursuit of conceptual understanding, procedural skills and fluency, and application with equal intensity—is needed to help children gain a deep understanding of the content and make sense of relevant context. The balance of context and content requires you to be intentional when selecting tasks as you design lessons that enable your children to experience the wonder, joy, and beauty of mathematics. In this chapter's Reflection and Action, you will have an opportunity to analyze several units of instruction to determine the level of instructional balance in your mathematics experiences.

Engaging Children in Rich Problem-Solving Tasks

Every mathematics classroom uses tasks to engage children. A mathematical task is a mathematical problem or set of problems that focus on a related mathematical idea or concept (Smith & Stein, 1998; Stein et al., 2000). Tasks can vary in their levels of cognitive demand and fall into one of two categories: low-level tasks or high-level tasks. Low-level tasks focus on memorization and practical procedural computations; high-level tasks require students to think, reason, and make sense of mathematical ideas (Boston, 2012). Table 5.3 provides a brief description of high-level tasks and notes where they intersect with the practices we shared in Table 5.1 Foundation for Equitable Teaching.

Table 5.3 *Characteristics of Mathematical Instructional Tasks*

High-Level Tasks	Foundation for Equitable Teaching (from Table 5.1) Equity-Based Math Teaching Practices (EMT) Math Teaching Practices (MTP) Standards for Mathematical Practice (SMP)
Focus children's attention on the use of procedures for the purpose of developing deeper levels of understanding of mathematical concepts and ideas.	EMT Go deep with mathematics. MTP Build procedural fluency from conceptual understanding. SMP Reason abstractly and quantitatively. Attend to precision. Look for and express regularity in repeated reasoning.
Usually are represented in multiple ways, such as visual diagrams, manipulatives, symbols, and problem situations. Making connections among multiple representations helps develop meaning.	EMT Go deep with mathematics. MTP Use and connect mathematics representations. SMP Use appropriate tools strategically.

(Continued)

Require complex and non-algorithmic thinking—a predictable, well-rehearsed approach or pathway is not explicitly suggested by the task, task instructions, or a worked-out example.	EMT Leverage multiple mathematical competencies. MTP Support productive struggle in mathematics. SMP Make sense of problems and persevere in solving them. Reason abstractly and quantitatively.
Require children to explore and understand the nature of mathematical concepts, processes, or relationships.	EMT Go deep with mathematics. Leverage multiple mathematical competencies. MTP Establish mathematical goals to focus learning. Implement tasks that promote reasoning and problem-solving. Use and connect mathematics representations. SMP Make sense of problems and persevere in solving them. Reason abstractly and quantitatively. Look for and make use of structure.
Require children to access relevant knowledge and experiences and make appropriate use of them in working through the task.	EMT Leverage multiple mathematical competencies. Draw on multiple resources of knowledge (math, culture, language, family, community). MTP Use and connect mathematics representations. SMP Model with mathematics Use appropriate tools strategically. Attend to precision.
Require children to analyze the task and actively examine task constraints that may limit possible solution strategies and solutions.	EMT Go deep with mathematics. Challenge spaces of marginality. MTP Establish mathematical goals to focus learning. SMP Make sense of problems and persevere in solving them. Reason abstractly and quantitatively.

Source: Adapted from Smith and Stein (1998).

Although the Foundations for Equitable Teaching (see Table 5.1) do not stand in isolation, you might notice the ones listed below were not included in Table 5.3. Nevertheless, we believe that each is critical as you plan how to engage your children in a TMSJ classroom and should be intentionally included because they support discourse and student identity.

Equity-Based Mathematics Teaching Practices:

Affirm mathematics learners' identities.

Mathematics Teaching Practices:

Facilitate meaningful mathematics discourse.

Pose purposeful questions.

Elicit and use evidence of student thinking.

Standards for Mathematical Practice:

Construct viable arguments and critique the reasoning of others.

In the ideal TMSJ classroom, teachers must be intentional in centering rich problem-solving tasks with a high level of cognitive demand. Rich problem-solving tasks provide children the opportunity to engage in high-level thinking, and they encourage reasoning and access to mathematics through multiple entry points and foster problem-solving through varied solution strategies (NCTM, 2014). Rich problem-solving tasks, especially those that connect to real-life problems, are essential for TMSJ to ensure that children's mathematics experiences are meaningful and relevant:

- They provide opportunities for multiple entry points and multiple representations.
- They require higher-order cognitive effort.
- They enable children to build on the knowledge they bring into the classroom.
- They provide opportunities for collaboration.

As educators, every task we select and implement can enhance a child's mathematics experience or serve as a barrier to them experiencing mathematics. Pause and consider the experiences children receive in your classroom. Take a moment to complete the Check-In activity.

 # Check In

1. Do all children consistently engage in high-level thinking?
2. Are all children consistently exposed to tasks that offer them multiple problem-solving entry points?
3. Do the tasks consistently foster varied solution strategies?

If you answered *yes* to all three questions, consider how often your children also see their lived experiences and/or are exposed to other children's lived experiences in mathematical tasks. If you answered *no* to one or more questions, reflect on what you could do differently in the future to change your response to a yes.

Rich problem-solving tasks encourage children to engage in the process and showcase their knowledge and understanding. They do not lend themselves well to too much direct instruction or over-scaffolding tactics such as demonstrating procedures, using keywords, or taking over the thinking when a student is stuck. The key is to provide students the time and space to explore and grow their thinking skills, using appropriate discussion and collaboration, leveraging the knowledge in the room, and providing specific feedback in the moment. In Figure 5.2, let's compare how two similar tasks can engage children.

Figure 5.2   *Example of a Rich Problem-Solving Task*

Task A	Task B
35 + 57 =	Kyla is volunteering to collect signatures for a petition. In the first two weeks, she collected 57 signatures. After the third week she had collected a total of 92 signatures. How many signatures did Kyla collect in the third week?

If Task A were posed to children, they would either count on from thirty-five or count on from fifty-seven. In addition, children could use manipulatives to solve the task. Although there are a variety of ways that children could engage in the task, essentially the focus would be on adding two numbers for the sake of adding two numbers. For Task B, first, a teacher would need to ensure children understand the context of the task and what is happening in the task. Then, children can have the opportunity to independently make sense of the task. In contrast to Task A, Task B does not readily provide the operation children will engage in. In Task B, children will need to use strategies to make sense of the situation and decide which operation or operations will lead them to a solution.

You might be reading this book and thinking, "My children can't . . ." or "My children always struggle with. . . ." It is time to eliminate the deficit mindset when using rich problem-solving tasks with your children. Children are naturally inquisitive and creative thinkers. Think about how children demonstrate their thinking and abilities when they are unencumbered by classroom processes; for example, when children are outside during free time (e.g., recess or before school) and they are inventing games collaboratively, building different structures, imagining all kinds of stories and activities. In these moments you are seeing children's intuitive and inquisitive nature without curriculum materials, without directions, and without teacher guidance. Thus, we must provide them with opportunities to engage in critical, provocative thinking as well as opportunities to use manipulatives and explore. We must move beyond only valuing whether they can just "get the answer" to valuing whether they make sense of their problem-solving process. Give them ample opportunities to formulate ideas and thoughts.

Let's look at another example from middle school: Jonathon's Bullying Survey (Figure 5.3).

Figure 5.3   *Jonathon's Bullying Survey*

Jonathon is surveying students in his school about bullying. So far, he has surveyed 600 students and 20 of them responded that they have witnessed a student being bullied. What percentage of the students have witnessed bullying?

In this task, the children must unpack the word problem, make sense of what the task asks, and then use their problem-solving skills. In addition, rich problem-solving tasks provide opportunities for children to engage in discourse with their peers to discuss their chosen problem-solving method. Finally, rich problem-solving tasks provide an opportunity to introduce or respond to social issues that impact the children in your classroom.

USING PROBLEM-SOLVING TASKS BASED IN SOCIAL ISSUES

You might be wondering "What does a mathematics task rooted in social issues look like?" Let's illustrate using an important social issue: the lack of mathematics textbooks in a classroom.

Mr. Harris has 37 children in his class. Twenty-five mathematics textbooks are issued to a typical-size class. How many more mathematics textbooks does Mr. Harris need to ensure every child in his class has a book?

The task is based on the Grade 2 standard: fluently add and subtract within 100. One key facet of teaching mathematics for social justice is ensuring mathematics tasks are standards based, as it is essential that students learn foundational components of mathematics. In addition, this task explores a social issue directly impacting Mr. Harris's classroom—the lack of materials and an overcrowded classroom. As children complete the task, they will simultaneously begin to ask key questions:

▶ How many children should be in a classroom?

▶ Is this normal? Do other classrooms have this issue? Do other schools have this issue?

▶ Who can rectify this issue and how?

As we have mentioned, problem-solving tasks are geared toward developing children's deeper understanding of mathematics. They are usually word problems that offer a context that can help children see how mathematics is "real." Take a moment and reflect on this concept of "real" mathematics. Often, teachers tell children they are being prepared for the "real world" and that all mathematics is useful, and they will need it "someday," but this often doesn't connect with their lived experience or school. We invite you to reframe "real world" as "real life" and be intentional when selecting problem-solving tasks that allow children to experience real-life mathematics through applications that are related to them and are in a familiar context. This may include such contexts as the latest video game your students have said they play, their home lives, a school or community issue, a social issue, a relevant current event, or some combination of these. Table 5.4 provides examples of lesson topics that are featured in the Corwin Mathematics series *Mathematics Lessons to Explore, Understand, and Respond to Social Injustice* (2020-2022). You can view one per grade band in Appendix C as well as on the companion website at https://qrs.ly/wbfixtr. This series, along with several other books, provides lessons that extend beyond just "fun" mathematics tasks to engage children and focus on meaningful mathematics.

Table 5.4 *Real-Life Mathematics*

Real-Life Mathematics	Example TMSJ Lessons
Contexts focus on children's lived experiences and their day-to-day lives.	Feeding Ourselves and Others (Grades K-2)
	Playground Prejudice (Grades 3-5)
	The True Cost of That $29 T-Shirt in the Store Window* (Grades 6-8) qrs.ly/xnffo8u
	Cor(o)ner Stores and Food Apartheid (Grades 6-8)
	Listen to GLSEN* (HS) qrs.ly/5bffo92
	Do Just Some Students Take Honors Courses? (HS)
Reimagines the classroom mathematics experience by engaging children in mathematics learning experiences rooted in social issues.	Feeding Ourselves and Others (Grades K-2)
	The Value of a School Lunch (Grades 3-5)
	Challenge Ableist Assumptions in Mathematics Problems* (Grades 3-5) qrs.ly/lxffo9b
	Middle School Mathematics to Explore People Represented (Grades 6-8)
	Smoking and Vaping: Targeting of Marginalized Communities (Grades 6-8)
	Bringing Healthy Food Choices to the Desert (HS)

Real-Life Mathematics	Example TMSJ Lessons
Allows children to experience mathematics as a tool to investigate, understand, and seek solutions to problems that mean something to them.	Examining Air Quality* (Grades K-2) qrs.ly/n5ffo9q
	Journey for Justice: The Farmworkers' Movement (Grades K-2)
	Water Is Our Right, Water Is Our Responsibility (Grades 3-5)
	Single-Use Plastics (Grades 3-5)
	Gerrymandering of Voting Districts (Grades 6-8)
	The Mathematics of Toxic Air Emissions (Grades 6-8)
	Hurricanes and Weather (HS)
	iPhone Cost (HS)

 You can view or download the lessons indicated with asterisks at https://qrs.ly/wbfixtr

When children realize mathematics is applicable to their everyday lives, they see it as valuable and have a desire to engage. The key is to remember whatever the tasks are, they must be authentically child-centered. To ensure tasks are authentically child-centered, you must develop a relationship with the children to ascertain their interests (revisit Chapter 3). Issues can range from playground access to fair policing. Anything that impacts children's lives can be a social issue. The key differentiator in teaching mathematics for social justice is it must be a *justice* issue and not just an issue impacting children (e.g., video gaming, social media). At the heart of TMSJ is a focus on preparing students for experiences and life beyond the classroom.

When it comes to social issues, refrain from pre-selecting what you believe is the social issue impacting children. Provide children the space and opportunity to share what they feel are the social issues impacting them and then incorporate the social issue (or issues) into the context of a mathematics task. We do not recommend reinventing the wheel in your development or modification of rich tasks. Use this as an opportunity to remix existing material.

TRY THIS: RECOMMENDED BOOKS

For additional resources and guidance in creating mathematics tasks, check out the following books by Lou Edward Matthews, Shelly M. Jones, and Yolanda A. Parker.

- *Engaging in Culturally Relevant Math Tasks: Fostering Hope in the Elementary Classroom* (Corwin, 2022).
- *Engaging in Culturally Relevant Math Tasks: Fostering Hope in the Middle and High School Classroom* (Corwin 2023).

Both books take a deep dive into culturally relevant mathematics teaching and provide practical approaches for planning and creating culturally relevant mathematics tasks.

Opportunities to Showcase Children's Thinking

We have a strong belief that every child enters the classroom with some skills and abilities. Although these skills and abilities vary, it's our responsibility to help children build on them. Throughout the school day, children have moments to display their strengths and their interests, particularly when outside of the classroom. Have you noticed how children light up when they are not in the traditional classroom setting? Often, they are loud and boisterous when on the playground, in the cafeteria, in the courtyard, at an assembly or pep rally, at a school sports event, or on a field trip. During these times, they show off outfits they've chosen, art they've made, or tell a story about something amazing that happened. Their natural interests emerge and they are constantly showing off their creativity, their enthusiasm, and their skills. How can we harness this energy inside the classroom? How can children likewise showcase their creativity, their enthusiasm, their skills—and importantly, their thinking—in the classroom?

 **Check In**

Take a moment to list all the ways that children in your environment are encouraged to showcase their thinking.

List of ways to showcase thinking:

_____

_____

_____

_____

_____

_____

What kind of opportunities did you list? If you're like us, you may have listed things like a mini presentation, a visual creation, showcasing their musical talents, and/or teaching the class a unique problem-solving method. As we consider various ways children can showcase their thinking as individuals, in small groups, in whole groups, or some combination of these, the key is to provide children multiple opportunities. A TMSJ classroom is a free-thinking space that gives students ample room to attempt new things, flex their creative muscles, make mistakes, and grow. As previously discussed, it must be a safe space where children are free to be themselves while problem-solving and collaborating. Let's delve into the role of discourse for engaging children and showcasing their thinking.

ENGAGING CHILDREN THROUGH DISCOURSE

"Discourse is the purposeful exchange of ideas through classroom discussion, as well as through other forms of verbal, visual, and written communication" (NCTM, 2014, p. 29). After one has selected rich problem-solving tasks, discourse in the TMSJ classroom provides children the opportunity to further their engagement in a manner conducive to their identities. Your job is to create the optimal environment for discourse to occur. Here are three ways to actively engage children in meaningful discourse:

1. Provide children prompts and questions to get started with and sustain discourse.

2. Use children's thinking to lead and guide the discussions.

3. Facilitate a mathematics connection process as children engage in discourse.

THE ROLE OF QUESTIONS TO DEVELOP DISCOURSE

Questioning has long served as a tool to help clarify thinking and help folx better understand ideas and situations. In the classroom, children use questions to clarify their understanding, and educators use them to probe children's thinking. Consider the four primary types of questions: gathering information, probing thinking, making the mathematics visible, and encouraging reflection and justification (NCTM, 2014).

➤ **Gathering information:** Students recall facts, definitions, or procedures.

➤ **Probing thinking:** Students explain, elaborate, or clarify their thinking.

➤ **Making the mathematics visible:** Students discuss mathematical structures and make connections among mathematical ideas and relationships.

➤ **Encouraging reflection and justification:** Students reveal deeper understanding of their reasoning and actions.

An equitable mathematics environment uses questions as a key pillar to develop and enhance classroom discourse. Questioning provides everyone in the room the opportunity to further expand their knowledge and understanding of the content and context, especially when it involves a social justice issue.

BUILDING MY KNOWLEDGE BASE TO DEVELOP DISCOURSE

As a mathematics educator, you must move beyond "answer getting" to making sense of mathematics to create an equitable mathematics classroom environment. Questions should be posed from the children's and teacher's perspective in an environment that is naturally inquisitive. Using a social justice lens, one seeks to challenge the status quo, leading one to question everything. Think about one of the most challenging questions you have encountered as a teacher related to classroom mathematics content, context, or both. How did you respond? What could have been done differently? In a child-centered environment that encourages children to showcase their thinking, collaborate with others, and ask questions naturally, there will be times when children pose questions you do not know the answer to. That is okay, as it is not expected that you are the bearer of all knowledge. However, you likely honor children's questions and thoughts. One way to honor them is to seek ways to research the answer collectively when you do not know the answer. The key is to be honest with yourself and the children as it relates to your knowledge and understanding, and then continue to learn.

For a teacher to prepare for the variety of questions from the children, they must be a continual learner. For many reasons, most teachers have gone through similar educational experiences that provided them with partial truths and were created to maintain the status quo. In addition, most teachers have not been challenged to think critically regarding their content area of expertise (NCTM, 2023). Consider how often teachers of specific content are asked to teach outside of the assigned context area. This creates a need to learn about and seek information for other content areas continually. Learning about the content standards is not enough. There are three areas one should focus on when building one's knowledge base: history, content, and identity (see Table 5.5).

Table 5.5 *Building Your Knowledge Base*

Topic	Description	Suggestions
History	One must gain a historical understanding of how we got here; in addition, history should be viewed through the lens of historically excluded groups.	• Read books. • Engage in discussions of community histories. • Have discussions with likeminded folx.
Content	As when engaging in teaching mathematics to create an equitable mathematics environment, one must have a depth of content knowledge.	• Develop a depth of understanding of the standards. • Read the common core progression documents.
Identity	Often, one is only aware of one's lived experience. It's also necessary to explore and understand others' lived experience.	• Review your identity markers periodically. • Research all you can find regarding identity markers.

Focusing on these three areas provides a solid foundation for a knowledge base, and we encourage you to not do this in isolation. You need to identify colleagues who believe in collaboration and can provide varying perspectives to collaboratively engage in building your respective knowledge bases. When seeking colleagues, focus on quality, not quantity. The goal of working together is to collaborate and impact children's lives, not just become a large "social club." In Chapter 9, Building a Community of Collaborators, we further explore the importance of collaboration with colleagues.

Engaging Children Beyond the Classroom

In Chapter 4, we wrote about the importance of providing collaborative opportunities to prepare children to apply mathematics and the skills and competencies they learn in your room to life beyond the classroom. This also means it is important to ensure that they get ample and consistent practice using these skills and competencies outside of your classroom. Table 5.6 offers some strategies for doing this.

Table 5.6   *Beyond the Classroom Strategies*

> - Have children do counting exercises outside of the classroom (e.g., when walking to different parts of the school; home activities).
> - Involve families in children's mathematics activities.
> - Create a community garden or a community recycling program that incorporates mathematics.
> - Engage in community forums using mathematics as the foundation.
> - Consider two-part lessons in which Part 1 starts inside the classroom and Part 2 must be completed outside the classroom.
> - Embed community components within mathematics activities.
> - Integrate children's out-of-school activities (e.g., sports, drama or theater, music, jobs, or family or social activities) into mathematics activities that enable them to make sense of mathematics in their environments.

The key to extending mathematics experiences beyond the classroom is to do just that: Have children engage in activities that begin in the classroom with a plan to help them see ways they can connect the learning to their lives, school, or community.

TRY THIS: BEYOND THE CLASSROOM

Elementary Activity	Middle School Activity	High School Activity
At the school playground, have children count the number of playground activities available for them to engage in. Allow them to discuss.	Have children determine the number of playgrounds in their community. Allow them to discuss what equipment is missing from their playground and determine the cost to add the equipment. Allow children to determine the boundaries of the community and understand it may vary from child to child based on their lived experience.	Have children review the number of playgrounds in the community and determine the cost to build a playground. In addition, connect with the health teacher, ELA teacher, or both, to have children write about the benefits of community playgrounds.

These examples are intended to simply provide ideas for activities beyond the classroom. For any activity, the goal is to make it collaborative and relevant to your children. Where possible, educators planning in vertical teams can design activities in which children can make connections to mathematical ideas learned in earlier grades.

To hear more from the authors about the social justice mathematics teaching framework, listen to this **conversation with Dr. Childs and Dr. Staley.**

qrs.ly/epffuls

Summary

The importance of engaging children in a social justice mathematics classroom extends beyond the integration of social justice topics into mathematics lessons. It shapes how you intentionally design the mathematics environment, set expectations, and engage your children.

Engagement, the fourth cornerstone, sets the foundation for TMSJ classrooms grounded in an equitable and engaging mathematical environment. In this chapter, we discussed the Foundations for Equitable Mathematics teaching, the need for a balance of context and content, as and a balanced instructional approach for content. Our conversations then focused on the rich problem-solving tasks and the need to showcase children's thinking, which elevated the role of discourse and importance of questions.

 # REFLECT

Respond to the following question.

- What does engagement look like, feel like, and sound like?

_____

_____

_____

_____

_____

_____

_____

_____

_____

_____

_____

 # ACT

1. Add your key takeaways and next steps to your TMSJ Action Plan.

2. Use the following steps to determine the dominant instructional approach in your classroom.

   a. Select a unit of study to analyze.

   b. Identify the primary instructional approach for each lesson.

   c. Record the totals (number of lessons or instructional days) for each pedagogical approach.

   d. Calculate the percentage for each approach by dividing the number of lessons (days) by the total number of lessons (days).

   e. Use the % of lessons column to create a stacked bar graph.

   f. Repeat steps a–e for at least 2 additional units.

3. Which of the three pedagogical approaches (procedural, conceptual, or problem-solving) does your instructional plan lend itself to?

4. Where are the opportunities to shift pedagogical approaches within a unit to establish a more balanced approach?

5. What are some of the underlying characteristics of units that appear to be "out of balance"?

Sample Chart: Instructional Approach Continuum (Where Are You?)

Pedagogical Approaches	Unit: _____		Unit: _____		Unit: _____	
	# of lessons (days)	% of lessons (days)	# of lessons (days)	% of lessons (days)	# of lessons (days)	% of lessons (days)
Procedural						
Conceptual						
Problem-solving						
Total number of lessons (or instructional days)						

online resources Available for download at https://qrs.ly/wbfixtr

_____

_____

_____

_____

_____

_____

_____

_____

_____

Where to Next?

The importance of engaging children in a social justice mathematics classroom extends beyond the integration of issues into mathematics lessons. It shapes how you intentionally design the mathematics environment, set expectations, and engage your children. Instructional practices, tasks chosen, and role of discourse are all within your purview, but they are often influenced by factors outside of your classroom that are beyond your control. In the next chapter, we will take a look at the importance of having allies inside and outside of the school community.

PART 3

BRINGING SOCIAL JUSTICE ISSUES INTO YOUR CLASSROOM

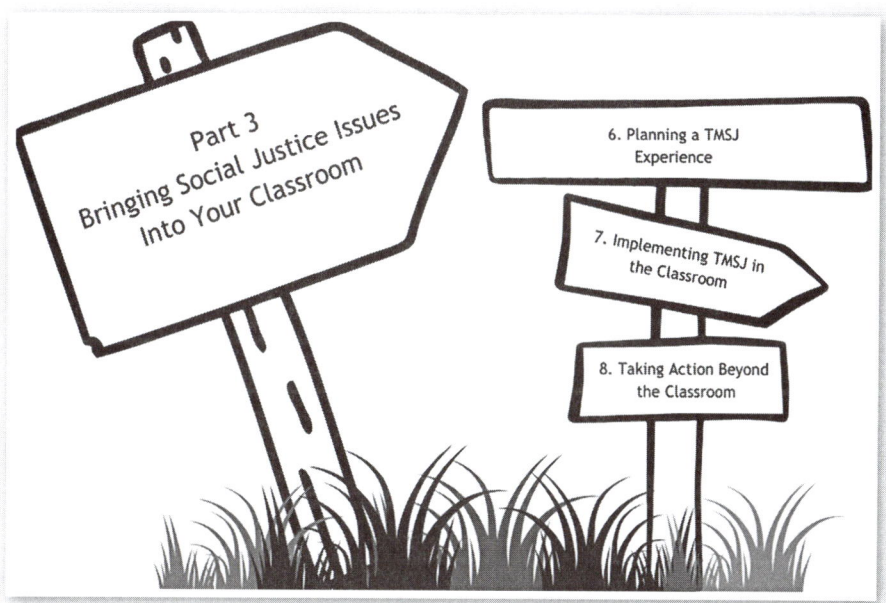

Now that you have had the opportunity to understand who you are in this work (Part 1) and explore an effective TMSJ classroom (Part 2), it is time in Part 3 to provide you with specific steps to bring TMSJ to life. In this core portion of the journey, you will be provided with the tools needed to create a social justice mathematics experience during your mathematics instructional time. Be prepared in this part to explore and reflect on how to bring social justice issues into your classroom (Chapter 6), how to implement a social justice mathematics lesson (Chapter 7), and how to take action beyond the classroom (Chapter 8).

6

PLANNING A TMSJ EXPERIENCE

In previous chapters, you have spent time reflecting on your identity, better understanding the children in your environment, and evaluating your instruction with an eye toward creating a classroom that is engaging, participatory, and holds all children to rigorous learning standards. Exploring each topic establishes the foundation for understanding how to plan and enact a mathematics lesson geared toward social justice and designed to close the achievement gap. In this chapter, we will walk you through how a TMSJ lesson comes together and provide the tools needed to develop an all-encompassing TMSJ experience. Specifically, you will

▶ learn how to brainstorm and select social justice issues to explore for a TMSJ experience,

▶ identify the keys to planning for social justice contexts,

▶ learn how to best support children in the TMSJ classroom, and

▶ make sense of strategies for bringing social justice issues into your classroom.

Brainstorming and Selecting a Social Justice Issue for a TMSJ Experience

One of the first steps when planning a TMSJ learning experience is to think about the kinds of social justice issues that your children value. Provide them time and space to showcase their thinking. For a TMSJ lesson, it is not enough to contextualize tasks in ways that are relevant or interesting to children; it needs to be an opportunity to use mathematical exploration to both learn mathematics and solve a social issue children find worthy of solving. Let's look at some things to consider in this selection process.

DIFFERENTIATING A RELEVANT TASK AND A SOCIAL JUSTICE TASK

A relevant learning experience can be anything children identify as interesting or relatable to them: a hobby, sport, or other fun activity they enjoy engaging in; how they spent their weekend; or something they read in the news or on social media. When a task is relevant, it speaks to the child's interest, which may or may not directly relate to a social justice issue. By contrast, a task with a social justice focus indicates a socially impactful outcome. A socially impactful outcome provides children the opportunity to engage in mathematics that either directly impacts them (e.g., becoming financially literate) or impacts their community (e.g., learning the benefits of a recycling program through data analysis). Both relevant and social justice-specific tasks have value, can provide children with rich mathematics experiences, and can be engaging. Complete Check In: Social Justice Topics to explore several topics and categorize them as simply a relevant issue, a social justice issue, or both.

 **Check In**

Social Justice Topics

Topic	Is it relevant, social justice-focused, or both?
School recycling	
Latest video game	
Building a community playground	
Gerrymandering	
Bullying	

WHAT SOCIAL JUSTICE ISSUES ARE APPROPRIATE FOR MY CHILDREN?

The key to selecting applicable social justice issues to bring into your lessons is first to identify what your children experience and what they care about. If children are old enough to experience injustice, they are old enough to make sense of it through the lens of mathematics and make sense of a pathway forward. It is essential, then, to collaborate with children to determine the best issues to address through a TMSJ lesson. Before discussing social justice issues with children, take some time to list some social justice issues from your own lens.

If children are old enough to experience injustice, they are old enough to make sense of it through the lens of mathematics and make sense of a pathway forward.

Check In

Social Justice Issues

List social justice issues that you believe are relevant to the community and/or would be interesting to the children under your purview. In addition, include a brief rationale for listing the issue.

Issue	Rationale

online resources ⏳ Available for download at https://qrs.ly/wbfixtr

In a TMSJ experience, one must always remain cognizant of the mathematics classroom setting, as topic appropriateness will vary from setting to setting, and social justice issues will vary when children themselves identify them. As an educator, you must tactfully determine the extent of the discussion. Understanding an issue may be uncomfortable for you but normal for them; however, as the educator, you ultimately have the final decision on whether the issue will be selected. Use discretion when determining the topic's appropriateness, especially if you are new to TMSJ. For example, you may be in a community in which access to healthcare is a widely discussed topic; however, a fellow TMSJ educator in a different community may be unable to broach the topic of access to healthcare. This is fine, as the goal is to provide children access to meaningful social justice topics, even if the topics vary from community to community. There is no one-size-fits-all and no single correct approach.

HOW DO SOCIAL JUSTICE TASKS HELP CHILDREN FIND MEANING IN MATHEMATICS?

As mentioned before, every mathematics educator at some point in their career has heard children ask, "Why do I need to learn this stuff?" That is an understandable and valid question. Thanks to technology, children have the world literally at their fingertips with microcomputers in their pockets and access to unlimited information, so when we hear this question, we must refrain from offering superficial responses. TMSJ provides the opportunity to make natural connections and show the significance of mathematical content. It makes room to explore mathematics conceptually rather than just procedurally. It provides the opportunity to see how mathematics makes sense and can be applied to actual problems. It enables children to grow and make sense of their worlds by caring about their communities and finding ways to help others. The question of "Why do I need to learn this?" becomes moot because inherent within a social justice task, children can see and understand the "why."

HOW DO I COLLABORATE WITH MY CHILDREN AND THEIR CAREGIVERS TO IDENTIFY SOCIAL JUSTICE TOPICS?

As mentioned, collaboration is critical when working with children to identify social justice issues they care about. In Chapter 3, we discussed ways to get to know your children's interests, likes, and dislikes, but how do we take it a step further to learn what social issues matter to them, their families, and their communities? First, remind children of things they have previously expressed liking, disliking, or interest in. Be specific so that children know you have heard and seen them and that you care about what matters to them. Next, start a conversation about social justice issues. Explain in your own words and in age-appropriate terms what social justice means. For younger children, you might talk about fairness—a concept they readily understand. For older children, you might frame it in more academic terms as previously described—social justice means equal rights for all peoples and the possibility for everyone, without discrimination, to benefit from economic and social progress worldwide. Offer a few examples, using some from Table 6.1 if you like, and ask what issues your children, their families, friends, and community members care about.

Table 6.1   *Social Justice Topics*

School boundaries	Immigration	Library access
Allowances and budgets	Fair policing	Diversity of educators
Educational materials in primary language	Quality housing	Food insecurity
Technology access	Food deserts	Protecting wildlife

Keep the conversation open-ended, as you want to get them talking at first. There are no right or wrong answers here. Write down everything students say where they can see it (e.g., on a whiteboard or poster paper) *exactly how they say it,* which captures and honors their thinking. Ask probing questions to get details so you can paint a picture of each issue children care about.

For example, a child mentions they want to address having more snacks available during lunch. Through probing questions, you can ask why this is important, what the pros and cons are, and what the possible solutions are. You may then determine through intentionally probing questions the root of the issue: Is the child hungry and does the child feel they are not getting served enough food? You can then craft an experience around the issue of food insecurity and lean into the data that 1 in 6 children reside in food-insecure households (Rabbitt et al., 2023). This exploration allows children to express what they value and feel needs to be addressed in their community. You will have no shortage of issues to consider as you and your children brainstorm topics. As you select issues, you and others will naturally start to have questions concerning the topics, and you can question their understanding of the topic to learn what they already know and want to find out. Always remember, one of the keys to TMSJ is letting issues organically come from those most proximate to the children and their families.

Next, survey families to ascertain their input. This can be done via a communication home. See Figure 6.1 for a sample letter.

Figure 6.1 *Family Letter About Math Class*

Dear Families,

We are beginning to explore teaching mathematics for social justice. The United Nations defines *social justice* as equal rights for all peoples and the possibility for everyone, without discrimination, to benefit from economic and social progress worldwide. This means we will use standards-based mathematics in our classroom to explore issues in the community your children care about and want to influence (e.g., recycling, food security, financial literacy, safe communities). This will help your child see how mathematics is a useful tool in their lives, and it will help us identify social issues in the community that we can become involved in to effect change. To ensure these mathematical experiences are meaningful for your child and beneficial to their learning, please list the top three social justice issues you feel impact your community. Although we will not be able to explore everyone's stated social justice issue, the goal is to determine a theme among the issues and select one or more throughout the year that are overarching and will best impact the community.

1. _____
   _____

2. _____
   _____

3. _____
   _____

online resources Available for download at https://qrs.ly/wbfixtr

Once families have completed this task, compile their feedback and integrate it with the ideas your children came up with to determine the best issue to address initially. As stated in the letter, you cannot address all of the issues; however, you can earnestly capture them and, when incorporating an issue, authentically unpack it using mathematics in a meaningful way.

Planning for the Social Justice Context Level

We recognize that not every educator, student, or community will be ready to dive into the deep end of using mathematics to explore social justice issues. Some issues may be viewed as controversial or sensitive, and we know that folx enter TMSJ with different perspectives and at different levels of readiness. Our goal in this work has always been to identify entry points for every educator, student, and classroom, to help you have the greatest success in teaching mathematics for social justice, so we propose a way of organizing possible contexts into levels: beginning, intermediate, and advanced. We define *context* as the circumstances that form the setting for an event, statement, or idea, and how it can be fully understood. Table 6.2 walks you through the nuances of each level, describing the nature and intensity of the issue, the amount of pushback you might anticipate for each level, and example topics. The three levels, which are not grade-level specific, help you identify and delineate your environments so you can plan accordingly to implement a social justice mathematics lesson. Remember the topics must cater to the environment, be engaging for the children, and add value to the community and families. To that end, although one topic may be appropriate for one grade level in a specific community, it does not automatically translate to the same grade level in a different community. Thus, you need to have a keen understanding of your community.

Table 6.2 is not an exhaustive list of topics. However, it provides a solid foundation to give folx an idea of the multitude of contexts one can use to engage children in teaching mathematics for social justice.

We invite you to revisit the list of issues you have gathered from your children and families earlier in this chapter (Check In: Social Justice Issues) and determine the social justice context level for each. As you select the issue and begin planning to integrate it into a lesson, remember that the focus of TMSJ is to show children how they can use mathematics to understand the issues and ultimately take action to have an impact in their communities. In the next section, we will discuss supporting children in understanding the context and the content.

Table 6.2   *Social Justice Levels and a Sample List of Social Justices Topics*

Beginning Topics (Level 1)	Intermediate Topics (Level 2)	Advanced Topics (Level 3)
Unlikely to receive pushback from families due to the nature of the topic; will need to simultaneously educate both the children and families.	*Possible to receive pushback from families due to the nature of the topic; will need to educate both the students and families simultaneously.*	*Likely to receive pushback from families due to the nature of the topic; will need to educate both the students and families simultaneously.*
• Access to healthy food • Community garden • Cost of living • Library access • Playground access • Recycling • Wages (minimum)	• Access to rich educational experiences • Body image • Bullying • Clean water • Insurance rates • Insurance (risk pools) • Public transportation • School boundaries • School uniform and access to clothes • Wages (gender)	• Climate change • Drug abuse • Equitable housing access • Fracking • Gender equality • Gender identity • Gerrymandering • Healthcare • Immigration • Incarceration • Forms of protest • Natural disasters and response • Police brutality • Suicide • Taxes • U.S. budget spending (i.e., education, military, etc.)

Supporting Children in the TMSJ Classroom

To give students the best chance at success in a TMSJ classroom, the proper supports must be in place before introducing social issues. You may need to take some steps in advance to help students get accustomed to such supports *before* you layer in the social justice topics:

1. Ensure all students receive regular access to grade-level mathematics content.

2. Familiarize students with technology supports—giving them time to learn and apply them—that will become second nature once they are engaging in TMSJ lessons.

3. Plan for adequate time and space for a full TMSJ engagement.

ACCESS TO GRADE-LEVEL CONTENT

Support starts with ensuring that all the children receive grade-level mathematics content. As described in Chapter 5, TMSJ does not sacrifice the rigor of the content for the experience. When implementing a social justice mathematics lesson that is grounded in a rich problem-solving experience and rich mathematical discourse, the number one pushback is, "What if children are not on grade level?" Although this is a real situation in many—if not most—classrooms, we must shift how it is addressed. Too often this is addressed with a focus on remediation for specific students through structured or unstructured interventions. Such interventions often replace core instructional time, which precludes children from receiving grade-level material and encourages a chronic cycle for some children of perpetually needing to "catch up." This is often a reason educators hesitate to integrate social justice experiences into the mathematics classroom.

There must be a mindset shift about how to work both with children who need support to access grade-level content and those who have a stronger grasp of grade-level content. In the first case, we must reconsider the notion that it's the children who must be fixed. It's time to change the systemic structures and practices that subject students to nonstop interventions and pull-outs and view them through a deficit lens. Interrupting a systemic "remediation focus" begins as you plan for the current series of lessons, unit, or a social justice mathematics lesson by identifying which children need support with accessing grade-level content on a given mathematical topic. Use Try This: Supporting Children With Access to Grade-Level Content to help you make the shift.

> Supports starts with ensuring that all the children receive grade-level mathematics content.

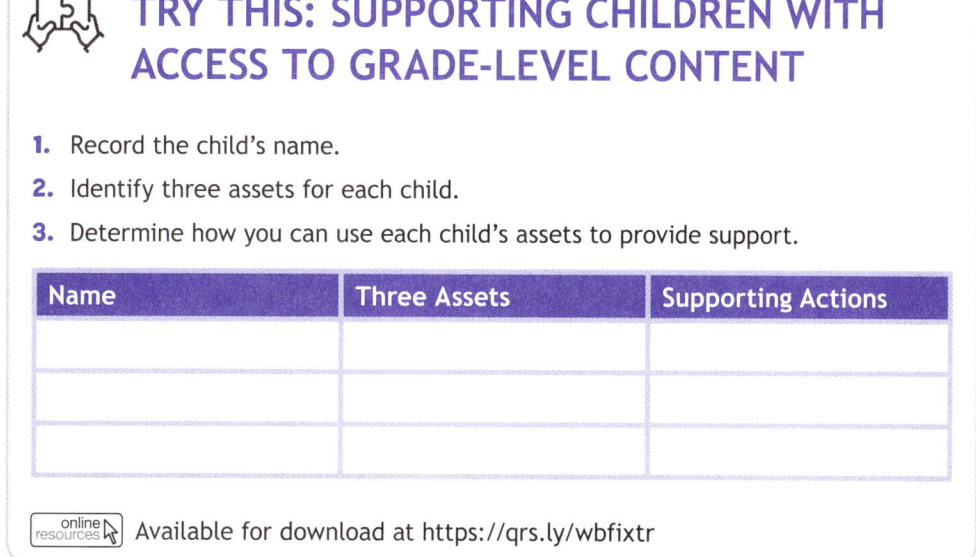

TRY THIS: SUPPORTING CHILDREN WITH ACCESS TO GRADE-LEVEL CONTENT

1. Record the child's name.
2. Identify three assets for each child.
3. Determine how you can use each child's assets to provide support.

Name	Three Assets	Supporting Actions

online resources ⬆ Available for download at https://qrs.ly/wbfixtr

The key is to purposefully look at children through an asset-based lens when planning for instruction. As you completed Try This: Supporting Children With Access to Grade-Level Content, you probably noticed how viewing the assets of children requires a focus. TMSJ calls for us to be intentional in our actions to improve the mathematics experience for all children, especially the historically excluded ones. As educators, we must always first consider what children bring into the classroom environment. This asset-based mindset changes how we, as educators, provide children with educational experiences. There is less focus on what children cannot do and a keen focus on what children can do. We must understand every child enters the classroom with "something," and it is incumbent on us to build on that something, continue to cultivate their assets, and value those assets as we provide mathematics experiences.

Conversely, often folx mistakenly believe that when children are *on* grade level, the next logical step is to provide them instruction *beyond* their grade level. This thinking can lead educators to believe TMSJ is not for them; however, TMSJ is designed for *all* children. Remember that equitable mathematics instruction involves ensuring every child can engage in grade-level mathematics by allowing

- ▶ multiple entry points,

- ▶ opportunities for exploration and discourse, and

- ▶ varied solution approaches.

When children are *on* grade level, rather than racing them ahead to next year's content, the focus should be on going deeper with the mathematics content and context, helping them to truly master the application of mathematical thinking in multiple contexts. You can help these children go deeper by asking them to justify their solutions, further engage with their classmates, and allow them to showcase various ways of thinking (Aguirre et al., 2013). Children working on grade level will benefit from TMSJ as they have time to truly unpack the social issue contexts and consider ways to impact their communities.

TECHNOLOGY SUPPORTS CONTENT AND CONTEXT

The second step of support involves the intentional use of technology, which will enable children to grow a deeper understanding of the mathematics content and further develop an understanding of the social justice issue. Technology is a tool that today's children have always had access to and will always have access to. Thus, it is imperative that we harness this power and use it to emphasize multiple mathematical representations and explore more authentic and realistic contexts (National Council of Teachers of Mathematics, 2023).

Technology enables children to engage in mathematics experiences unavailable to previous generations. Children can explore mathematical topics with the click of a button, further expanding their curiosity, willingness, and desire to understand the mathematics content they are engaging in in-depth. In addition, social justice issues can be complex topics. Children can use technology to research the social justice issue in-depth, from multiple lenses, to expand their knowledge base. This exploration allows them to think beyond and see mathematics beyond the static textbook page. Table 6.3 provides some ways to harness the power of technology to support a TMSJ classroom.

Table 6.3 *Harnessing the Power of Technology to Support the TMSJ Classroom*

	Benefits of Integrating Technology	Examples of Technology
Mathematics Content	• Increases opportunities for authentic learning experiences that promote learners' success • Provides opportunities to increase inclusivity and access to the mathematics	Online scientific/graphing calculator to enhance the problem-solving experience and allow exploration opportunities Internet to research social justice issues
Social Justice Context	• Allows deep exploration of the social justice context by providing a tool to research the context and a tool to make conjectures regarding different scenarios connected to the mathematics content • Provides a tool to allow for reflective practice and progress monitoring • Allows children to use familiar tools in situations directly connected to education • Allows children to learn to use technologies they will use beyond the classroom	Digital spreadsheets to input and analyze data Online presentation software to showcase collaboratively identified solutions

Source: Adapted from National Council of Teachers of Mathematics (2023).

MAKING SPACE FOR SUPPORT

If ensuring children are focused on grade-level content is the first step of support, and offering access to useful technology is the second, the third step is to ensure you are creating time and space for children to learn and make sense of the context and the mathematics. Creating this space means intentionally and effectively using instructional time to take deep dives in the mathematics content and social justice issue and tap into children's innate inquisitive nature. This means recognizing you may have to go beyond a traditional one-mathematical-topic-a-day approach. A lesson may extend over multiple days or class periods. It takes time to unpack the social justice issue; work through the mathematics of analyzing and understanding the issue; consolidate the mathematics learned; consider, propose, and enact solutions to the issue; and demonstrate how those solutions might have a positive impact on the community. However, recognize that in a problem-solving classroom, a single TMSJ experience may be able to hit multiple mathematical standards, so in the amount of time it takes to engage in a TMSJ experience, you are likely to cover as much mathematics, and in more depth, than in a traditional one-topic-a-day format.

Check In

Take a moment to consider the structure of your current classroom setting. Do children have opportunities for deep exploration?

If yes: What are some things that you have done to cultivate these opportunities?

If no: What needs to be done to provide this type of environment? What are your barriers and how might you get beyond them?

Once children are getting the necessary support to experience on-grade mathematics through problem-solving tasks rich with discussion, once technology supports are enabled, and once you have planned for the necessary time and space to explore a topic in depth, you can now begin to layer in social justice issues.

Bringing the Social Justice Issue Into Your Classroom

Remember, in a TMSJ experience, the mathematics content remains a foundational part of the lesson and it is the context that changes. Some mathematical topics do not lend themselves to TMSJ experiences, and that is okay. Don't force it. In the early stages, the goal is to slowly modify current traditional contexts to ones with social justice issues based on what children have identified as meaningful to them. Take time to review Table 6.2 and the social justice issues indicated. Consider your environment and which topics, as discussed with the children and community, would best benefit them and which topics may not be as meaningful to them. We recommend that you view incorporating TMSJ with a unit or semester thematic lens as it will be highly challenging to make sense of a unique social justice issue lesson by lesson. The goal of TMSJ is depth, not breadth, of social justice issue coverage. Depending on your knowledge of the social justice issue, you will learn and grow just as the children learn and grow. Be patient with yourself as you begin to reimagine the mathematics experience and incorporate meaningful topics that can not only positively change a child's life but also impact their community in a meaningful way.

When incorporating social justice issues, feel free to take baby steps in the beginning and work to ensure a solid meaningful overall experience. Think of a snowball rolling down a hill. At first, it may seem like a small and possibly meaningless starting point, but just as a snowball rolls downhill and grows in size, your experience of TMSJ will grow incrementally, layer by layer, and improve over time. The trick to this process is not to invent or design a new task or tasks to align with the content you are about to teach, but to take existing curriculum tasks and modify the contexts. If you are currently using a quality curriculum, the tasks have already been vetted. Thus, you can easily modify the context, knowing that the mathematics content is solid. When modifying the tasks, work to ensure the mathematics makes sense based on a realistic context. We don't want to inadvertently contrive a context to suit the mathematics in a way that doesn't make sense and further confuses children. For example, let's say you want to incorporate an environmental justice lens based on input from the children and their families, with a keen focus on clean water. However, the current mathematical topic is box and whisker plots. The goal is not to use made-up data or force fit the issue into the mathematical topic, as this will further confuse the children and possibly confuse you. If this happens, take a moment to reset and reflect. Use this as a learning opportunity and determine the best pathway forward; it may be engaging in the same social justice issue and it may require reconsidering the social justice issue. Remember to not sacrifice the quality and rigor of the mathematics content.

The secret to a successful TMSJ experience is having students engage in rich discourse, as described in Chapter 5, and supporting conversation by asking thought-provoking questions. These questions enable children to expand their mathematical thinking and their worldview. Thought-provoking questions go beyond the normative "What answer did you get?" or "What is your solution?" They focus on the child's problem-solving process, thinking pathways, and thoughts about the social justice issue you are exploring. In a TMSJ experience, children must understand the mathematics content and social justice issues. We should inspire children to do more and challenge them to think and reflect, so we must ask forward-thinking, thought-provoking questions to help

> Be patient with yourself as you begin to reimagine the mathematics experience and incorporate meaningful topics that can not only positively change a child's life but also impact their community in a meaningful way.

children critically think about the value and meaning of mathematics. Here are some sample questions:

- ▶ I'm interested in understanding your thinking. Can you tell me about your problem-solving process?

- ▶ Show me how you arrived at your solution. What steps did you take?

- ▶ Can you think of any other ways of solving the problem?

- ▶ How does the math in this task help you understand the social justice issue better?

- ▶ What questions do you have regarding the social justice issue?

- ▶ What are some things you believe you can do to address the social justice issue?

- ▶ What would you like to learn further about the social justice issue?

Consider these questions as conversation starters, but for mathematics settings and geared for children. Essentially, your goal is to provoke the children to share their thinking in a manner comfortable to them. In addition, as they respond, you are provided opportunities to get to know them, understand them, and develop a relationship with them. Initially, some children may hesitate to respond and seek to identify the "right words to say." Use this to create a safe space and environment and reiterate that all thoughts are valid and valued. The more children share in the setting, the healthier the setting becomes and the more inclined the children will be to share and engage more.

Summary

You may be asking, "How many social justice issues should one select?" The key for TMSJ is focusing on quality, not quantity. Over the school year, many only engage in two to four social justice topics. Think of the issues as overarching themes that intertwine over time. TMSJ is not a one-and-done experience. Thus, as you begin to select issues, consider which issues can be explored for extended periods. This will enable you to focus on depth instead of breadth when engaging in TMSJ.

As you have learned, when planning a TMSJ experience, one must consider multiple components; however, the goal is not to master all of the components at once. But over time, work to get better in each component consistently. Getting better not only takes time but also it is helpful to have some like-minded colleagues to collaborate with. In Chapter 8, we will discuss building a community of collaborators. For now, take time to focus on ways to begin this work with what you currently have.

To hear more from the authors about their reflections about planning a TMSJ experience environment, listen to this **conversation with Dr. Childs and Dr. Staley.**

qrs.ly/y3ffulu

REFLECT

As you plan to implement a social justice mathematics lesson, you must pause and analyze your environment—classroom, school setting, and community. Next, take some time to review the social justice context levels and the social justice issues you are considering for the lesson.

- Which social justice context level is most appropriate for your environment?

- What steps will you take to ensure that your children and families are ready?

_____

_____

_____

_____

_____

_____

_____

_____

_____

ACT

1. Add your key takeaways and next steps to your TMSJ Action Plan.

2. Discuss community concerns with the children and collectively brainstorm possible solutions.

Community Concern	Possible Solutions

3. Complete the Planning TMSJ Status Check table by assessing your current understanding of the planning for TMSJ experience components, and record your next steps to improve your knowledge.

_____

_____

_____

_____

_____

Planning TMSJ Status Check

Component	Self-Assessment*	Actionable Next Steps	Notes
Mathematics standards			
Selecting rich problem-solving tasks			
Collaborating with children			
Building your knowledge base			
Incorporating social justice issues			
Asking thought-provoking questions			

* Rate your understanding of each component as 1, 2, 3, or 4 according to the following scale:

1: Not yet; just getting started
2: Getting it; progressing in a way that is likely to result in success
3: On track; skilled in the component
4: Achieving; continuing to develop and improve my skills and knowledge

 Available for download at https://qrs.ly/wbfixtr

Where to Next?

Now that you have unpacked the planning of a TMSJ experience, it is time to learn how to implement and support TMSJ in the classroom. The next chapter is the most crucial chapter in the book as now you will have the opportunity to apply your learnings and give the children a memorable mathematics experience featuring a social justice issue.

IMPLEMENTING TMSJ IN THE CLASSROOM

7

The foundation has been laid, and the time has come for you to teach a social justice mathematics lesson. Remember that you have done the work to learn about your children (Chapter 3); create a mathematics environment that centers your children's culture, fosters a sense of community, and encourages collaboration (Chapter 4); engage all children so that they can reach their potential (Chapter 5); and identify social issues that are relevant to your children (Chapter 6). In this chapter, we will

- ▶ walk through the implementation of a social justice mathematics lesson,

- ▶ share examples of social justice mathematics lessons, and

- ▶ unpack the core lesson components through the framework for social justice mathematics lesson.

Implementing a Social Justice Mathematics Lesson

Now that you have explored how to bring social justice issues into the classroom (see Chapter 6), it is time to move to the next stop on the journey: implementing a social justice mathematics lesson. A reminder before we begin: As you keep the work moving forward, remain cognizant of your mindset throughout the TMSJ experience, as you will constantly be challenged internally and externally when engaging in the work. TMSJ educators are dedicated to the work of facilitating change through the use of mathematics, which requires a mindset centered on five critical components: enriching experiences for children, impactful contexts, collaborative opportunities, action oriented, and maintaining a positive outlook. Table 7.1 lists these mindset components, their rationale, and tips for their enactment.

As you keep the work moving forward, remain cognizant of your mindset throughout the TMSJ experience, as you will constantly be challenged internally and externally when engaging in the work.

Table 7.1   *Mindset When Implementing a Social Justice Mathematics Lesson*

Mindset Component	Rationale	Enactment Tips
Enriching experiences	TMSJ enhances mathematics content through enriching contexts. Children should be excited to engage in TMSJ experiences.	Develop goals. Never sacrifice the rigor of the mathematics content.
Regularly incorporating impactful contexts	TMSJ is designed to provide children with impactful experiences. Children are able to see the application of the mathematics they are learning.	Maintain a focus on the bigger picture of the social justice context and mathematics content. Seek cross-curricular connections.
Collaborative opportunities	TMSJ has a keen focus on collaborations between children, between children and educators, and, as you will learn later, collaborations among educators. Children work with others who may have different backgrounds and perspectives.	Provide ample opportunities for children to discuss mathematics and social justice issues. Provide ample opportunities for the children to work together to make sense of varying problems and solutions.
Action oriented	TMSJ has an expectation that children take some action or create a product to share with others.	Seek children's ideas for sharing with others. Be prepared with suggestions to get thinking started. Ensure children connect the social justice context and mathematics when sharing with others.
Positive outlook	Mathematics lessons are never perfect, and TMSJ experiences are no exception, thus the need to have the mindset of always using every experience as an opportunity to learn something new and move forward.	Keep the main thing the main thing when leading TMSJ experiences. Be bold.

Reflecting on the mindset described in Table 7.1 will help you as you begin to implement a TMSJ experience. Experiences first start with a lesson, so in this chapter we will take a deep dive into some sample lessons and engage in some reflection. We include one lesson from the Corwin Mathematics series *Mathematics Lessons to Explore, Understand, and Respond to Social Injustice* (2020-2022) for each grade band: Grades K-2; Grades 3-5; Grades 6-8; and Grades HS. You can find these complete lessons in Appendix C or access them online through the links or QR codes in the Check In box below. Select two lessons to review and complete the following Check In. In the next section, we will take a closer look at the different components of a social justice mathematics lesson.

 ## Check In

- What is the lesson's mathematical content focus?
- What is the social justice issue, and why is it important for children to make sense of the issue?
- What common/different instructional components are embedded in each lesson?

Grades K-2	**Grades 3-5**
Lesson 5.13	**Lesson 5.1**
Early Elementary Mathematics to Explore People Represented in Our World and Community	**Families Matter**
(Number and Operations)	(Fraction Concepts and Operations)
Written by Courtney Koestler, Eva Thanheiser, Mary Candace Raygoza, Jeff Craig, and Lynette Guzmán	Written by Nicky Meindl
 qrs.ly/2mffoa2	 qrs.ly/dkffoa8
Source: Koestler et al. (2022). *Early elementary mathematics lessons to explore, understand, and respond to social injustice.* Corwin.	Source: Bartell et al. (2022). *Upper elementary mathematics lessons to explore, understand, and respond to social injustice.* Corwin.
Grades 6-8	**High School**
Lesson 9.4	**Lesson 7.5**
How Many Meals Can Minimum Wage Buy?	**Humanizing the Immigration Debate**
(Statistics and Probability)	(Statistics and Probability)
Written by Elizabeth O. Ayisi and Colleen Carman	Written by Ayensur Ozturk and Steve Lewis
 qrs.ly/lfffoab	 qrs.ly/keffoac
Source: Conway et al. (2022). *Middle school mathematics lessons to explore, understand, and respond to social injustice.* Corwin.	Source: Berry et al. (2020). *High school mathematics lessons to explore, understand, and respond to social injustice.* Corwin.

Although the selected lessons each intentionally have a different focus, each one shows how TMSJ experiences can encompass a range of topics, engage children in meaningful learning—contextually and mathematically—and invite them to become a change agent. Hopefully, these lessons help you envision your children engaging in a social justice mathematics lesson.

Social Justice Mathematics Lesson Components

As you review these lessons, one similarity you will find among them is that they all have an introduction or launch, some type of exploration or investigation, and a consolidation and closure. Let's look at some specific sections from these lessons to unpack the components of a social justice mathematics lesson and situate each component in the framework for social justice mathematics lessons (see Figure 7.1) shared in *High School Mathematics Lessons to Explore, Understand, and Respond to Social Injustice* (Berry et al., 2020).

Figure 7.1   *Framework for Social Justice Mathematics Lessons*

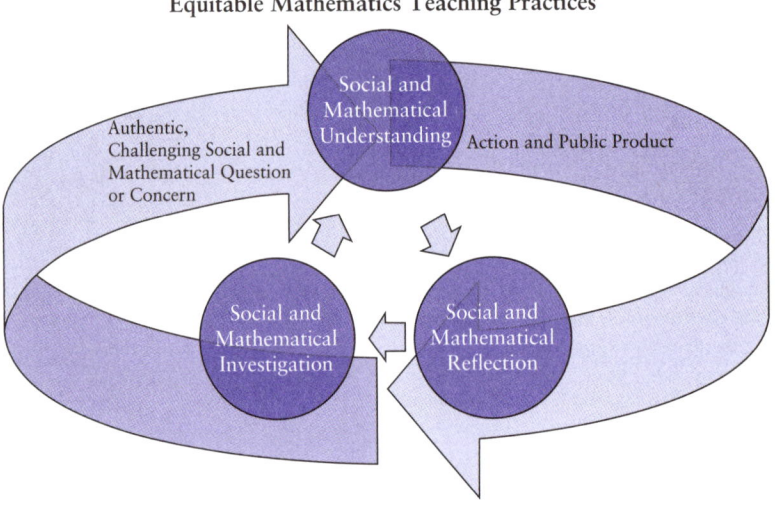

Source: Berry et al. (2020).

INTRODUCTION

The beginning of any social justice mathematics lesson (SJML) serves as an opportunity to connect to children's interests, activate prior knowledge—mathematics, lived experience, funds of knowledge—and create a meaningful and sometimes personal purpose for learning mathematics. SJMLs focus on an *authentic, challenging, social and mathematical question or concern*. Social justice contexts—especially those that arise from your children and families—can help children to observe patterns, critique information, learn to ask questions, and reflect (see Figures 7.2 through 7.5).

Figure 7.2 Early Elementary Launch: Mathematics to Explore People Represented in Our World and Community

It is important to set the stage for lessons that have a social justice context. ⋯⋯⋯

Opening routines that build community help children to value diverse identities. ⋯⋯⋯

Introduce issues with children's literature/video and question prompts. ⋯⋯⋯

LESSON FACILITATION

Day 1

Greeting (10 minutes)

- Begin the lesson with an opening circle greeting and community building activity entitled, *I love my neighbor who . . .* or *I have solidarity with people who . . .* The purpose of this activity is (1) to support children to get to know one another better and to build trust and community and (2) to challenge assumptions about one another and identify similarities and differences with one another.

Launch (15 minutes)

- Introduce the idea of shrinking the world's population down to a village of 100 people that represents it by reading the first few pages of *If the World Were a Village* by David J. Smith. Use a globe or map to provide context if necessary. (Supplementary websites or videos can also be used; see Teacher Resource 1.)

- After reading the introduction, ask:

  + *In the village of 100 people, how many people do you think would live in the United States [or North America]?*

  + *How many people do you think would speak English?*

- You may also engage children in wondering about or predicting other characteristics of the world's village discussed in the book (for example, the ages of the residents of the village, how many residents have regular access to electricity, or other aspects that you think the children would find interesting).

Source: Koestler et al. (2022).

Figure 7.3 Upper Elementary Launch: Families Matter

LESSON 1 FACILITATION

What Makes a Family?

Launch (20 minutes)

- Begin with a think–pair–share. Ask students to answer these questions with a quick-write or drawing:

  + *What makes a family?*

  + *What does a family look like?*

  + *Who is in your family?*

- Next, have students share their responses in small groups. Walk around the classroom during these discussions to note what common and unique ideas arise among the students.

TEACHER NOTE

You should consider the social identities of the students in your classroom, such as if students are adopted or in foster care, when discussing families. Many times, these students have no idea when, or even if, they will see their birth families, so it is important to remind them of what makes a family and what ways their foster/adopted guardians care for them like "blood family." Find and read books throughout the year that show the foster care system to give students mirrors to see themselves in as well as windows for others to learn about a piece of their experiences.

LESSON 2

Launch (10–15 minutes)

- Begin with this question: *How can we find out what types of families are represented in the books most read by students?* Allow students to share their ideas.

- Display the following goal on the board: "Gather and report on data of family structure representation in the most checked-out library books."

Provide opportunities for children to respond in writing/drawing before sharing their thoughts with others.

Awareness of children's social identities allows you to plan learning experiences throughout the year for children to see themselves.

Have children share ideas for completing the lesson task(s) to build self-confidence and promote a sense of agency.

Establish goals (mathematics and social justice) to focus the lesson.

Source: Bartell et al. (2022).

Figure 7.4   Middle School Launch: How Many Meals Can Minimum Wage Buy?

Providing questions with a social justice context that requires mathematics to answer focuses the lesson.

Launch (10 minutes)

- Distribute Worksheet 1 (*Minimum Wage and Cost of Living*).
- Spark curiosity by showing the video "When Can You Afford More Big Macs? In 1968 or Now?" (https://bit.ly/3rx4geE) and have students jot down some initial thoughts and observations on the worksheet (question 1.a).

Source: Conway et al. (2022).

Figure 7.5   High School Launch: Humanizing the Immigration Debate

For topics that are political and possibly emotionally difficult, make sure to remain neutral and encourage children to share from diverse perspectives.

Act 1: "Who Is Dayani Cristal?" (90 minutes)

Students explore the immigration debate learning more about Dayani Cristal, a Mexican immigrant wanting to cross the border illegally. This lesson is a powerful starting point for studies on US immigration, as well as domestic issues influenced by undocumented immigrants.

Take steps to prepare children and families for this lesson. (Social Justice Context Level 2)

+ *Why do you think people migrate?*
+ *How does socioeconomic class affect people's choices or abilities to achieve their goals?*

Source: Berry et al. (2020).

EXPLORATION

The heart of the lesson is to develop your children's *social and mathematical understanding* through a rich *social and mathematical investigation*. Understanding requires us to carefully balance the **what** (mathematics content), the **how** (mathematics practices and processes), and the **why** (a response to the social justice issue) as we

- ❯ facilitate the lesson,
- ❯ provide support, as needed, to help children access the mathematics and social justice context, and
- ❯ navigate discussions that invite and value your children's diverse perspectives.

The social and mathematical investigation serves as a catalyst for creating understanding and needs to be grounded in a rich problem-solving task and based on grade-level mathematics that are appropriate for exploring the social context. The tasks should build upon your children's own reasoning and problem-solving (see Figures 7.6 and 7.7).

Figure 7.6   Upper Elementary Exploration: Family Matters

Provide guidance as children explore so that they are attentive to the intersection of the social justice context and mathematical content. ....

- Tell the class that they will be looking closely at the pictures and mentions of families in the books for both primary and secondary characters. Tell the students that as they work, they need to come up with categories for the different types of families that they see in their books. Emphasize that they need the following for each category:
  - + A definition or description (for example, "two moms" or "includes foster children")
  - + An example from one of the books they are reviewing

Explore (45–60 minutes)

Use questions to probe and push children's thinking. ....

- Have students work in groups to develop categories. Circulate and pose questions such as these:
  - + *What categories did you create?*
  - + *Who did you include in your categories?*
  - + *Who is missing from your categories?*
- When you feel groups have a rich enough set of categories, bring them together for a whole-class discussion.

Source: Bartell et al. (2022).

Figure 7.7   High School Exploration: Humanizing the Immigration Debate

LESSON 2 FACILITATION

Act 2: Problem Creation and Looking at Immigration Data (90 minutes)

Tasks involving data allow children to create their own questions, building personal connections to make the learning more meaningful.

The goal of this portion of the lesson is to synthesize the highly contextualized information from earlier. Students begin to develop the ability to examine data and represent it in multiple ways.

- Allow time for students to start up digital devices to explore a website.
- Introduce Arizona OpenGIS Initiative for Deceased Migrants, humaneborders.info, and show how to do research on migrant mortality in different categories (e.g., year, cause of death, county of death, gender).
- Place students into groups of four.
- Have each group create a statistical question that requires research on migrants' deaths. For example, while one of the groups works on the question "How many male migrants die because of exposure in a given ten-year period?," another other group might work on the question "In which region of Arizona desert do people mostly die?"

Technology provides opportunities for children to explore the social justice context and mathematical content (support needed).

Source: Berry et al. (2020).

CLOSURE

Lessons often end with some type of summary and assessment of what children have learned. When implementing a SJML, it is important to focus on the mathematics your students have learned so that the mathematical learning outcomes are clear and remain in the forefront. It is equally critical that you make time for *social and mathematical reflection* and *action and public product.* Berry et al. (2020) share that the "questions posed by teacher and peers, and discussion about solution methods or ideas with peers . . . promote reflection—about the mathematics, the social issue, and how the two inform one another" (p. 61). The one major difference between SJMLs and other mathematics lessons is the opportunity for children to take action or develop a public product.

> The one major difference between SJMLs and other mathematics lessons is the opportunity for children to take action or develop a public product.

One result of investigating a social injustice is often a deeper understanding and awareness that somehow connects to

➤ identity—how we view ourselves;

➤ diversity—how we view others and their perspectives; and

➤ justice—how we view fairness and unfairness, unequal power relations, and the impact of bias. (Berry et al., 2020, p. 61)

You may have observed in each of the lessons previously reviewed there is a "Taking Action" section (see Figures 7.8 through 7.10). Having a list of possible actions to share with your children will help get the class thinking, and you will find that it will be more meaningful if your children develop their own actions.

Figure 7.8   *Early Elementary Closure: Mathematics to Explore People Represented in Our World and Community*

TAKING ACTION

Individual, Class, or School

- After the lesson, children can create infographics to articulate what they have learned to their parents and caregivers about the diversity of the world to embody the Learning for Justice Standards (Diversity 8 and 10). They can use these infographics to compare their local communities to the greater world community (Diversity 7).

- Once children explore the diversity of their local context, they often are curious about resources and supports for different peoples. Teachers can engage children in using knowledge they gain to take action. For example, when children in one second- and third-grade classroom found out that there were a large number of children and families who spoke Spanish and Chinese but their school only had a Spanish Family Liaison support staff member working at the school, they wrote letters to the principal to find out why and to ask if there was any way to get funding for the Chinese-speaking families in the district.

Local Community, Organizers, or Organizations State, National, or Systems Level

- You may also support children to identify people (e.g., family members, politicians) or organizations (e.g., their own school) that may hold assumptions about who is represented in different spaces and to think about calling in or calling out those assumptions with data.

Communicating With Stakeholders

- Before teaching the lesson, you can let children's families know that you will be exploring the diversity in our world, both at the global and local levels. You can invite family members in to share their strengths, resources, and insights with the children. For example, if there are multilingual family members that would be willing to come in and share (e.g., by reading a book in a language other than English), invite them.

Children's individual and collective reflection can often lead to actions that extend beyond the classroom.

Identify, in advance, people children can connect with as resources for learning and advocates for action.

Source: Koestler et al. (2022).

Figure 7.9 Upper Elementary Closure: Family Matters

Ideas generated by children make the learning more relevant leading to actions and products carefully designed for sharing with respective stakeholders.

LESSON 4 FACILITATION

Connecting Research to Action

This lesson provides an opportunity for students to connect their research to action. You can decide if you want to have different students or groups to take different forms of action or if you want to plan a common action (such as a presentation and/or letter-writing campaign) that everyone will participate in.

Launch (10–15 minutes)

- Ask the class to brainstorm ideas for how they can take action based on the research they have done so far. Here are some possible ideas:
  + Running a donation drive for books with diverse representations of families
  + Sharing results on social media and soliciting suggestions for book recommendations
  + Sharing results with family members

+ Sharing results with and/or writing letters advocating for change to the librarian, principal, school board, parent/caregiver teacher association (PCTA), or other members of the community

TAKING ACTION

As part of the end of the lesson, the students will write letters to the school librarian, school principal, school board members, or school district representatives to advocate for the addition of books that reflect the diverse family structures in both their community and their world. The students will detail why adding these books are important, provide suggestions for one or two books to be added to the school's library, and explain their reasoning for selecting the book.

As an extension activity, the students could work with their local and school libraries to highlight books that show diverse family structures. The students would be working with their librarians to create short summaries or book reviews to help entice other readers to select, read, and then share the book with others in their community. This could also include small art projects to create visuals to display about the highlighted books with diverse family structures, which would help draw attention to these books.

Source: Bartell et al. (2022).

Figure 7.10 Middle School Closure: How Many Meals Can Minimum Wage Buy?

TAKING ACTION

Encourage students to take one of the following actions:

- Write a letter to their state representatives with their conclusions and justifications about affordability of goods and services over time. Persuade state representatives to support their claim about minimum wage in 2020.
- Make a poster with their findings and a call to action. Hang it up around their school or neighborhood.
- Educate others about cost of living and minimum wage.
- Speak at a city council meeting.

Actions and products can extend beyond self-reflection to sharing indirectly (e.g., posters, social media posts, letters) or directly (e.g., presentations, invited conversations) with others.

Source: Conway et al. (2022).

To hear more from the authors about their reflections about implementing TMSJ in the classroom, listen to this **conversation with Dr. Childs and Dr. Staley.**

qrs.ly/yqffulw

Summary

In this chapter, you learned the key components of implementing a social justice mathematics lesson in the classroom. It all starts with one's mindset; once one has the right mindset, one will be better equipped to handle the ebbs and flows of TMSJ. Practical examples were provided to give you an opportunity to review effective lessons and further think through your process for developing your TMSJ experience. We then unpacked the common components of lessons—Introduction, Exploration, Closure—situated in the framework of social justice for mathematics lessons.

REFLECT

Implementing TMSJ is the most impactful chapter related to the TMSJ experience. Take a moment to reflect on and answer the following questions:

- How has your mindset been impacted through reading this chapter?
- What are you compelled to do differently as you move forward to implementing a social justice mathematics lesson?

_____

_____

_____

_____

_____

_____

_____

 # ACT

1. Add your key takeaways and next steps to your TMSJ Action Plan.

2. Now, take a moment to complete the table below to prioritize the actions you will take immediately and in the near future to implement TMSJ in your environment.

_____

_____

_____

_____

_____

_____

_____

_____

My Implementation Next Steps (Prioritizing Your Work)

Next Steps	Specific Action You Will Take and Rationale	Notes
Immediate action(s)		
In the near future		

online resources 🔖 Available for download at https://qrs.ly/wbfixtr

Where to Next?

Now that we have unpacked implementing and supporting TMSJ in the classroom, it is time to unpack how to take action beyond the classroom. TMSJ seeks to expand the mathematics experience beyond the traditional four walls of the classroom and enable children to apply their knowledge to real-life applications as well as seek ways to engage their families and communities.

TAKING ACTION BEYOND THE CLASSROOM

In earlier chapters, you learned the foundational components of a TMSJ experience. Now, it is time to take TMSJ to the next level by considering how to take action beyond the classroom. In this chapter, we will explore the following:

❯ The benefits of taking action beyond the classroom

❯ The keys to engaging families and the community

❯ Effective ways to involve administrators

❯ Showcasing your impact

After reading this chapter, you will be prepared to take TMSJ experiences beyond the traditional classroom.

Experiences Beyond the Classroom

Integrating action and experiences beyond the classroom is one thing that differentiates our view of a TMSJ experience from others engaged in this work. We believe to have a true TMSJ experience, children must move beyond just using mathematics to investigate and make sense of an issue and apply their learnings to a real-life scenario designed to impact their community. As stated earlier, mathematics was not conceptualized for the purpose of having children just solve a bunch of math problems; it was conceptualized as a way to make sense of life and the world around us all. Through this lens, let's explore the importance of extending the experience beyond the classroom.

In Chapter 6, we described the importance of helping children answer the question "Why do I need to learn this stuff?" This is the essential question behind our desire to engage children in TMSJ experiences. Mathematics is a meaningful subject, and children should be able to connect the mathematics content they are learning to real life. Extending TMSJ has three main benefits:

1. Children experience and make sense of mathematics beyond a static textbook page.

2. Children have the opportunity to experience the application of mathematics to real-life scenarios.

3. Children have the opportunity to impact their communities.

First, mathematics is so much more than some static problems on a textbook page. Mathematics, as described earlier, is a rich, joyful, fascinating, and all-encompassing subject that should not be limited in any form. TMSJ provides teachers with avenues to expand the mathematics conversation beyond just routine tasks, word problems, and problem-solving. Second, TMSJ provides children the benefit of experiencing the applications of mathematics to real life. Children should not have to "imagine" applications of the mathematics they are learning, nor should they have to fabricate them or apply them to contrived situations that don't match their day-to-day realities. They should experience practical connections that will enable them to view mathematics as a subject useful in life. The more they experience mathematics as valuable, the more likely they will desire to further engage in mathematics. The third benefit is that children are provided opportunities to impact their communities. They acquire tools that will benefit their communities now and in the future. Ultimately, the goal of the standards being implemented in United States' schools, and those around the world, is to prepare children for their higher education and/or career endeavors, which in turn prepares them to become productive citizens. But first, let's discuss how to extend TMSJ experiences beyond the classroom.

Extending social justice mathematics lessons beyond the classroom must be intentionally and collaboratively planned. Social justice issues are issues for a reason, as they bring some inherent challenges. Therefore, they will not magically disappear because you address them in the classroom. The classroom is used to explore and make sense of the social issue and underlying mathematics concepts, and to extend beyond that, you need a plan. If this sounds challenging, remember that everyone has a responsibility to address the social justice issues that you and your children explore, so do not feel the onus is just on you and your students to solve everything yourselves. Rather, the main goal is to play *your role* in addressing the issue—acting locally and effecting what change you can. Your plan starts by thinking with the end in mind. What do you want to achieve as an extension beyond the classroom?

In educational settings, the goal is to prepare children for life and what happens after one finishes school. But think for a moment: How often, during children's kindergarten through high school experiences, do they engage in activities outside of the classroom that started with a mathematics lesson? Take a few moments to complete Check In: Extending Lessons Beyond the Math Classroom and reflect upon your current classroom and extension ideas.

> Children should experience practical connections that will enable them to view mathematics as a subject useful in life. The more they experience mathematics as valuable, the more likely they will desire to further engage in mathematics.

Check In

Extending Lessons Beyond the Math Classroom

List the opportunities children have in your school to engage in mathematics activities outside the mathematics classroom.

_____

_____

_____

_____

_____

_____

Are there current opportunities in the mathematics classroom that can extend to the school or community? If so, what are they?

_____

_____

_____

_____

_____

_____

If you have never extended lessons beyond the mathematics classroom, what ideas do you now have?

_____

_____

_____

_____

_____

_____

As you think about your end goal, here are five tips for planning lessons that extend beyond the classroom.

1. Think back to the social issues you collected from Chapter 6 and think about what connections to mathematics you can make. Brainstorm with your children what kind of solutions they would like to see, why, and what role they might play.

2. If your school allows, have mathematics class outside at a community site. This provides a different learning space and the opportunity to explore mathematics beyond the classroom.

3. Collaborate with colleagues and design joint mathematics experiences beyond the classroom.

4. Participate in existing school and/or community activities. Engage your children in using mathematics as a tool to understand the social context.

5. Use technology (e.g., virtual reality headsets, interactive tablets) to allow children to explore beyond the classroom and make sense of specific mathematics connections.

The next step in your plan involves engaging families and communities at the outset to make sure they are informed, on board, and can even participate.

Keys to Engaging Families and the Community

Engaging families in TMSJ began in Chapter 6, where we discussed inviting families to offer ideas for social issues that could be used in lessons. A desired outcome of TMSJ is to impact communities, which ultimately impacts families. At their core, families want their children to receive high-quality educational experiences. Some form of education has been a crucial aspect of all cultures throughout history, regardless of cultural background or geographical location. In present-day schooling, understanding the diverse backgrounds of families being served is crucial, as their experiences vary. Some have had excellent experiences with mathematics education, while others have not. Many have been taught using a procedural lens, irrespective of culture. Therefore, as children engage in a problem-solving environment focused on social justice and critical thinking, educators must not only develop the children's mathematical abilities but also engage their families. Teaching mathematics for social justice requires thorough communication with families, explaining not just the content but also its context and significance in their children's growth.

Families play a vital role in the educational experience. They can either enhance or hinder it (intentionally or unintentionally). Hence, it's crucial, even before the start of the academic journey, to involve families—from day zero, the day before day one. Gathering their input and considering their perspectives during planning is essential. When one is incorporating TMSJ, and specifically a social justice issue, into the mathematics classroom, families will naturally have questions and concerns. These will range from apprehension to curiosity to possible pushback. The key is being prepared to engage families sooner rather than later, offering the rationale for using a social issue as context in a math lesson, and describing some of the events children may be able to participate in that extend into the home or community. In Chapter 6, Figure 6.1. A Letter for Families, we shared a letter introducing social justice and inviting parents to help identify social issues in the community. This is your entry point to start involving families in the TMSJ classroom, and many families will have questions:

- What does a social issue have to do with the math my child is supposed to learn?

- Is focusing on social issues a distraction from my child learning math?

- How will my child benefit from TMSJ?

- Are you indoctrinating children?

- Why can't my child learn mathematics how I was taught?

When teaching a social justice mathematics lesson, the key is involving families from the beginning of the experience by sharing the intentions of the lesson in advance, informing them of discussion topics that will come up during the lesson, sharing the activities children may be doing as part of the lesson, and describing the benefits children will experience and the impact they will have as a result of learning about and acting on the social justice issue. Benefits align with the CASEL 5 competencies covered in Chapter 1. Children develop self-awareness through growing and learning about themselves as they experience TMSJ. They grow in self-management by engaging in Standards for Mathematical Practice 1—engaging in making sense of problems and persevering in solving them. They learn responsible decision-making when provided continuous opportunities to engage in rich problem-solving. Relationship skills are strengthened through endless collaboration opportunities. Social awareness is a benefit as children gain a greater understanding of their community and how they can play a role in positively impacting it. In addition, some tasks will naturally involve families directly, allowing them to see firsthand the benefits of TMSJ. Families can play the role of supporter, nurturer, collaborator, and contributor.

- **Supporter**: ensures their child completes the tasks beyond the classroom, such as engaging with a local political leader, interviewing a community elder, purchasing an item for a project, or canvassing their neighborhood.

- **Nurturer**: helps their child understand the application and rationale for what they are learning as it relates to the social issue.

- **Collaborator**: provides the teacher with information, resources, and community artifacts for the lesson based on their lived or community experiences.

- **Contributor**: participates in the lesson or activities beyond the classroom by sharing their expertise related to the lesson's social issue (e.g., as guest speaker, live or virtually) or organizing or participating in a related community outreach activity.

When families are involved from the outset, they grasp the importance of expanding the educational experience beyond the classroom. The goal of TMSJ is to impact communities, offering direct benefits to some families beyond the classroom setting. For instance, this might involve access to a community garden, increased financial literacy, or the initiation of a community watch program. Providing families with options to engage in the TMSJ experience is pivotal. Families come from diverse backgrounds and experiences, particularly regarding their own understanding of and experience with school, including their own instructional experiences (Ambroso et al., 2021). There's a good chance that the mathematical experience their child receives differs from what they experienced. Hence, there's an opportunity to educate both the child and the family simultaneously on social issues, fostering positive change within their community.

> When families are involved from the outset, they grasp the importance of expanding the educational experience beyond the classroom.

Keep in mind, too, that schools are part of communities, and some families have deep roots in these communities, offering unique insights and perspectives on community dynamics and desires. Teaching mathematics for social justice isn't a one-sided endeavor; it's a collaboration among all stakeholders. When crafting lessons, consider how families and community members can be integrated. Find where the mathematics content intersects with everyday activities, such as grocery shopping, managing utility bills, or balancing a bank account. Explore ways to involve families and community members in these connections to help them see mathematics beyond mere textbooks or worksheets. Even before diving into teaching mathematics for social justice, seek opportunities for families to engage in their children's mathematical learning experiences. Encourage involvement through math tasks involving parents' input, enabling families to discuss social issues impacting the community. Invite family and community members as guest speakers to share how they use mathematics in their professional careers or everyday life. This approach aims to demonstrate to children the various real-world applications of mathematics. Additionally, ensure thorough communication with families by informing them extensively about the classroom experience. An example family letter has been included in Appendix B to assist in initiating this process.

Once you have decided how you will inform and involve families and other community members, the next step of your plan is to involve administrators.

Effective Ways to Involve Administrators

Administrators' main role in extending beyond the classroom is playing the role of path clearer. In this role, they work to ensure everyone has optimal access and support beyond the classroom. This includes providing additional space within the school building, providing space outside of the school building, collaborating with community members for outreach, and acting as a liaison with entities beyond the school for partnerships and sponsorships. Administrators are intertwined with the process from the beginning and they know that at this point, the mathematical content understanding has been achieved and the understanding of the social justice issue has been achieved. Therefore, taking action beyond the classroom is a natural extension and further helps children solidify their understanding of the material. Listed below are four tips for involving administrators:

1. Let administrators know at the onset of your TMSJ intentions.

2. Discuss with them how the lesson will align with the standards.

3. Invite them to participate in a TMSJ lesson.

4. Seek their guidance when engaging community members.

The final element of your plan for extending beyond the classroom has to do with how you and your children will showcase and celebrate the work you've done.

Showcasing and Celebrating Your Impact

> TMSJ educators understand the importance of celebrating success and acknowledging their children's hard work and dedication because it encourages the children to strive for future success.

Often, in TMSJ one becomes so intensely focused on the work that it is easy to overlook the importance of pausing for a moment to celebrate success and allow children meaningful opportunity to showcase their hard work. TMSJ educators understand the importance of celebrating success and acknowledging their children's hard work and dedication because it encourages the children to strive for future success. In this work everyone can easily become engulfed and so passionate about the work it takes most of their time and little time is left for other efforts. However, pausing to reflect and expressing gratitude in multiple ways show children an additional benefit of engaging in these experiences. Celebrating success also encourages them to continue to engage in more TMSJ experiences as they see the value add not only to the community, but within themselves also. This, in turn, leads to children wanting to continue to improve and ensure the work continues.

Celebration alone is not enough. You must also craft opportunities for children to showcase their efforts—as a form of celebration—by presenting their solutions, sharing with their families, and interacting with the community to share and appreciate the impacts of their work. Every TMSJ experience should culminate with some form of a presentation in a manner that aligns with the child's preferred method of presenting, which can vary from a formal presentation to a written narrative to a video synopsis. Allow children the opportunity to be creative and showcase their talents. Remember, TMSJ is a collaborative effort;

give children the freedom to express themselves and provide opportunities for them to innovatively collaborate on presenting. To maintain consistent messaging, collaborate with your children to agree on one specific thing they would like to share with their families. These sharing opportunities provide another touch point with families to continue to involve them in the TMSJ experience. Likewise, communities are the ultimate recipient of the work of a TMSJ experience; therefore, identify reasonable and practical opportunities for the children to share the results of the TMSJ experience with the community.

To hear more from the authors about their reflections about taking action beyond the classroom, listen to this **conversation with Dr. Childs and Dr. Staley.**

qrs.ly/8rffuly

Summary

In this chapter, there has been a keen focus on TMSJ beyond the classroom, engaging families, and involving administrators. TMSJ experiences are collaborative efforts. Thus, everyone must work together to ensure the children have the maximum benefits of the experiences. In addition, as you are learning, the benefits extend beyond just the children; they also positively impact everyone involved in the experience.

REFLECT

It is time to ensure you are applying your learnings thus far. Reflect upon the following questions and your progress in understanding the TMSJ experience.

- How has your thinking evolved in teaching mathematics?
- What can be done to enhance relationships and communication efforts with families?
- What are you looking forward to doing next in your environment as it relates to TMSJ experiences?

_____

_____

_____

_____

_____

_____

_____

_____

_____

_____

 # ACT

1. Add your key takeaways and next steps to your TMSJ Action Plan.

2. Review the list of social issues received from your families in Chapter 6 and select one or two for possible use in a future mathematics lesson. Identify the people (family or community members) who are possible resources for the lesson.

_____

_____

_____

_____

_____

_____

_____

_____

_____

_____

_____

Resources for My Social Justice Lesson

Social Issue	Resource (Family or Community Member)

online resources Available for download at https://qrs.ly/wbfixtr

Where to Next?

Now that you have learned how to take action beyond the classroom, it is time to build a community of collaborators. As you have learned, TMSJ is not a solo endeavor; thus, let's explore strategies for building collaborations.

PART 4

SUPPORTING AND GROWING YOUR PRACTICE

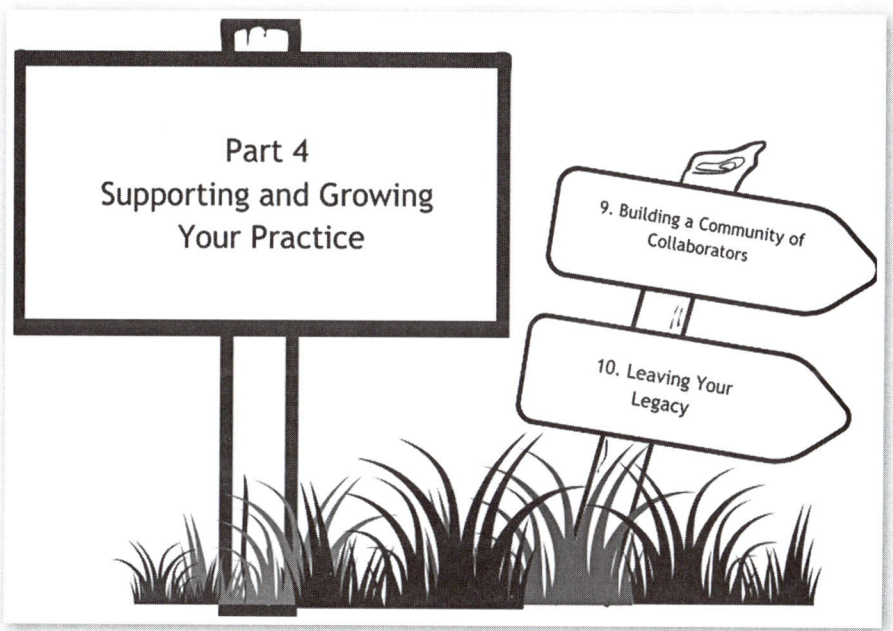

Part 4
Supporting and Growing
Your Practice

9. Building a Community of Collaborators

10. Leaving Your Legacy

You have come a long way on your journey, but we're not quite at our destination. It is now time to summarize what you have learned in Parts 1, 2, and 3 and identify what is needed in terms of support so you can engage in the work of TMSJ for the long haul. In Part 4, the two main focus areas are building a community of collaborators (Chapter 9) and leaving your legacy (Chapter 10). Being an effective TMSJ educator is a collective endeavor and an endeavor that seeks to live in perpetuity. Let's explore taking your work as a TMSJ educator to the next level.

9 BUILDING A COMMUNITY OF COLLABORATORS

TMSJ, as you have learned, touches on many aspects of your children's mathematics experiences, from teacher-to-child and child-to-child interactions; family- and community-member engagement; and classroom-, school-, and community-action-oriented opportunities. All of these experiences speak to TMSJ as a collaborative endeavor, which involves children, families, teachers, administrators, and community members, just to name a few. The key to designing successful mathematical experiences for your children is to build a community of collaborators who can support you along the journey by

1. contributing words of wisdom, advice, and guidance that help you navigate teaching with, about, and for social justice;

2. sharing insights that inform your current work or shape your plans; and

3. walking beside you to lift up the work and carry some of the load.

In this chapter, we will

➤ explore the roles of various collaborators,

➤ learn how to build *your* core group of collaborators, and

➤ identify the keys to growing a community of collaborators.

Let's first discuss what we mean by a community of collaborators, including how we distinguish between quality and quantity of collaboration and how such a community can contribute to the larger goal of TMSJ.

Who Is in Your Community of Collaborators?

As we've discussed, TMSJ is a rich and ambitious form of mathematics teaching. It cannot be done in a vacuum and is best not done alone, if possible. To get the most out of the effort for your students and yourself, and to increase your likelihood of success, it's important to establish a community of collaborators for purposes of support, guidance, and action. This community includes children, families, educators, and other community members (e.g., business owners, first responders, service workers, civic leaders and volunteers, clergy, athletes, community organizers) outside of the school building. Before imagining how to bring such a community together, it is essential to first understand the various roles collaborators might play in TMSJ and how they can support you in understanding your children, creating a TMSJ mathematical environment, increasing student engagement, and integrating social justice issues into your mathematics lessons. Here is an example of a few roles:

▶ **Advocate:** supporter who champions the need for and inclusion of TMSJ experiences.

▶ **Beneficiaries:** individuals or groups who receive a benefit from the TMSJ experience and resulting actions.

▶ **Collaborator:** provides the teacher with information, resources, and community artifacts for the lesson based on their lived or community experiences.

▶ **Contributor:** participates in the lesson or activities beyond the classroom by sharing their expertise related to the lesson's social issue (e.g., guest speaker, live or virtually) or organizing or participating in a related community outreach activity.

▶ **Guide:** offers suggestions and support for the children and, in some cases, the classroom teacher.

▶ **Leader:** supports the facilitation of the social justice mathematics lesson and action activities.

▶ **Learner:** shows an interest in how mathematics can be used as a tool to understand a social justice issue.

▶ **Supporter:** expresses an interest in the math lesson, social justice topic, and opportunities to engage with others.

▶ **Thought Partner:** thinks through various aspects of the lesson and extended activities in collaboration with the teacher.

Table 9.1 provides a list of constituents, what roles they each play, and how involvement benefits them and you.

Table 9.1 *Collaborators Play Many Roles*

Constituent Group	Role vis-à-vis TMSJ	As part of your TMSJ community, this group can
Children	• Learner • Contributor • Leader • Beneficiaries	• Be exposed to and learn to acknowledge and respect diverse perspectives from peers. • Deeply learn mathematics. • Have the opportunity to make an impact on the community.
Families	• Supporter • Contributor • Guide • Leader • Beneficiaries	• Be more involved, informed, and active participants in their children's education. • Offer their knowledge, skills, and interests to TMSJ experiences. • Share ideas, advice, or knowledge to incorporate into TMSJ lessons. • Learn mathematics content in a manner that may differ from their previous learnings. • See the enactment of social justice mathematics lessons. • Feel a sense of pride in their children's deep learning, empathy building, and community involvement. • Potentially benefit from community actions their children take.
Other Educators, Colleagues, & Administrators	• Supporter • Guide • Contributor • Leader • Advocate • Thought Partner	• Be a trusted critical friend to act as a sounding board and accountability partner. • Be a thought partner to brainstorm lesson ideas. • Advise on navigating dynamics of relationships among constituents. • Help scout other collaborators or needed resources from the community. • Benefit from cross-curricular lesson planning and enactment. • Learn from your model of integrating social justice into your topic area. • Support you if conflicts or questions arise.

Constituent Group	Role vis-à-vis TMSJ	As part of your TMSJ community, this group can
Outside Community Members	• Supporter • Guide • Contributor • Leader • Advocate • Thought Partner	• Be exposed to and learn to acknowledge and respect diverse perspectives from peers. • Have the opportunity to make an impact on the community. • Be directly involved in the local school and add cohesion to the community. • Offer their knowledge, skills, and interests to TMSJ experiences. • Share ideas, advice, or knowledge to incorporate into TMSJ lessons. • See the enactment of social justice mathematics lessons. • Feel a sense of pride in local children's deep learning, empathy building relationships, and community involvement. • Potentially benefit from community actions children take.

As you consider the different roles, think about the needs and benefits for each group of constituents. Being strategic and thoughtful in the partnerships you form can deeply enhance your TMSJ journey, which in turn benefits your children.

QUALITY VS. QUANTITY COLLABORATIONS

As a TMSJ educator, you understand the importance of time, which is a limited commodity for pretty much everyone. This means you also understand the need to focus on the quality of your collaborations and not the quantity of the collaborations you are engaging in or seeking to establish. As stated in earlier chapters, your TMSJ journey is a marathon and not a sprint, thus the need to intentionally select people who will support the needs you identify. Since you do not have time to waste, you do not want to spend time with too many educators who do not understand the value of TMSJ or who do not want to engage the children under their purview in TMSJ experiences. In other words, you need allies. It does not behoove you to have numerous folx to collaborate with if none of them are seeking the same things you are. They will feel like dead weight that you are dragging along, which will ultimately tire you out. Instead, focus on quality collaborators who have the same zeal you do and who have an equally strong desire to partner and improve. Here are several questions to ask potential collaborators. If they answer "no," reconsider partnering with them at this time in your TMSJ journey.

1. Do you believe every child deserves and should have access to a high-quality education?

2. Do you have a genuine passion for engaging children in rich and meaningful learning experiences that involve community and social issues?

3. Do you believe the work you are supporting can positively impact the children and their community?

4. Do you prioritize fostering long-term, meaningful collaborations over engaging in numerous short-term partnerships?

5. Are you committed to understanding and implementing the principles of TMSJ in your practices? (For educators only)

These questions will help you find collaborators who share your commitment and passion for TMSJ. They enable you to gain a glimpse into the mindset of the potential collaborators so you can begin to determine *who* to collaborate with and for *what* purpose. They increase the likelihood that your collaborations will center on shared objectives and quality interactions. They will help you feel confident that your collaborators will be there for you when you need them. We now turn our focus to the importance of establishing a collective understanding of the goal of TMSJ.

COLLECTIVELY UNDERSTANDING THE MORE SIGNIFICANT GOAL

When building a community of collaborators, you must maintain a continual focus on two main things: commitment to the bigger goal and sustainability.

When building a community of collaborators, you must maintain a continual focus on two main things: commitment to the bigger goal and sustainability. The bigger goal (to impact communities positively) is important because you and your collaborators need to be going in the same direction and seeking similar outcomes. This enables you to hold each other accountable as you strive to reach the ultimate goal. In addition, through meaningful collaboration, you can build upon each other's ideas as they relate to attaining your goal. This also allows you to start thinking through the importance of sustainability, as meaningful change requires a lasting effort. TMSJ experiences can not only change the community but also the status quo for society, so you need to plan for collaboration for the long term, as one cannot afford to get weary on the journey or stop short of the outcome. Often, folx enter this work with a short-sighted mindset, which ultimately limits the TMSJ experience. Maintaining a focus on quality allows TMSJ educators the latitude to scale with intentionality for long-term endeavors and not just to fill immediate needs. The more committed an educator is to TMSJ, the more opportunities they have to grow everyone's knowledge, create cross-classroom collaborations, and create sustainable systems that can stand the test of time. The bigger goal is too important not to work relentlessly and tirelessly until it is achieved.

Building a Core Group of Collaborators

We want to focus the rest of this chapter on building collaboration within one particular group of constituents: other educators. We have explored in other chapters ways to bring on board and involve family members (Chapters 3, 6, and 8), educational leaders (Chapter 8), and outside community members (Chapter 8), but other educational colleagues are *central* to your collaboration team, and therefore we want to spend a little more time here. Building a core group of education professionals to collaborate with consists of two goals: identifying like-minded educators to team up with and making time for collaboration.

IDENTIFYING A LIKE-MINDED T.E.A.M. OF EDUCATORS TO COLLABORATE WITH

T.E.A.M. means Together Everyone Achieves More. The work of TMSJ is a collaborative effort, thus the need to collaborate with educators who are like-minded. A T.E.A.M. mentality is key as you consider which educators to collaborate and form relationships with along your journey. We encourage you to pause and truly think about your specific needs. Take a few moments to complete Check In: In Search Of and reflect on your areas of need.

 **Check In**

In Search Of

Briefly describe how you would want a colleague to "show up" to do the following:

Support you in monitoring and checking your biases:

_____

_____

_____

Provide critical constructive feedback on your classroom setting and interactions with children:

_____

_____

_____

Walk with you as you learn about your children and school community:

_____

_____

_____

Design lessons that integrate school, community, and social justice issues into the classroom/school:

_____

_____

_____

Complement your specific skills and knowledge with their own:

_____

_____

_____

online resources ⟍ Available for download at https://qrs.ly/wbfixtr

Reflecting on your responses to Check In: In Search Of, different names may come to mind for each question, and that is okay. Each question brings to light a different need for you and a different level of vulnerability. It's important that you proceed knowing that educators who have a TMSJ mindset understand

▶ the power mathematics plays in the lives of their children inside and outside of the classroom, school, and community;

▶ that they have a role in developing each child's self-confidence as doers, creators, and thinkers of mathematics;

▶ that working with colleagues and walking with those who are on a TMSJ journey is a responsibility that should not be taken lightly;

▶ that together everyone achieves more. This means having a T.E.A.M. work mindset by understanding that everyone brings unique strengths, skills, and knowledge to the table and complements everyone else.

This mindset sets TMSJ educators apart from other educators because they know that collaborating on a TMSJ experience allows everyone to go to greater depths and reach greater heights than anyone attempting this journey on their own. When seeking like-minded educators, consider Table 9.2, which lists the key attributes of a TMSJ educator and what to look for when seeking a TMSJ educator.

Table 9.2   *TMSJ Educator Attributes*

Attribute	Look For
Has mathematics teacher knowledge	• Has a foundational understanding of mathematics content knowledge and pedagogical content knowledge • Continually seeks to improve content knowledge
Is collaborative	• Consistently seeks opportunities to partner with others • Creates a classroom environment based on collaboration
Is an open communicator	• Freely communicates with other TMSJ educators and stakeholders • Engages the children and their families in active dialogue regarding mathematics and social issues
Has community awareness	• Views the community through an asset-based lens • Seeks to understand the community from folx within the community
Believes in children	• Believes children can be successful in mathematics • Provides opportunities for children to showcase their thinking

When seeking like-minded educators, keep these key attributes and your responses to Check In: In Search Of at the forefront. These are baseline qualities to begin with and you should add attributes to the list over time.

SCHEDULE TIME FOR COLLABORATION

As a TMSJ educator, be willing to work creatively to find time for collaboration. In the work of TMSJ, time is one of the most precious resources, and TMSJ educators must be willing to create a schedule and make the appropriate time. A collaboration schedule is essential to give folx opportunities to work together consistently and meaningfully (Dufour et al., 2016). Frequent meetings give TMSJ educators opportunities to not only discuss mathematics and social issues but to bond and deepen their understanding of each other. Learning more about each other helps them to create more meaningful experiences for the children under their purview, as they can better understand each other's strengths and weaknesses and balance each other's skillsets. Set a meeting schedule that you can collectively stick to—at a minimum, once a month. The monthly convening ensures everyone has an opportunity to meet, share, and collaborate. In addition, with current technologies available, communities are not limited to face-to-face meetings and interactions. One can use video conferencing software, telephones, and social media apps to collaborate with folx around the world.

Growing a Community of Collaborators

A TMSJ educator is committed to developing a quality community of collaborators, which requires scaling the community, maintaining discretion, and committing to unapologetic support.

A TMSJ educator is committed to developing a quality community of collaborators, which requires scaling the community, maintaining discretion, and committing to unapologetic support.

First, intentional scaling is essential as a TMSJ educator. As you have learned, the work cannot be done individually, and it needs to be sustainable. The broader and more robust the community of collaborators, the more likely you are to have long-term success. As mentioned, this means not just looking for a lot of partners but also looking for high-quality, committed collaborators who can increase the overall impact of TMSJ, which enables the movement to grow in a more impactful manner. This entails doing outreach to other educators in the building beyond your grade-level team, other educators in the district, or other mathematics educators through social media groups or national organizations. Also consider researching the non-profit organizations in your local area for possible partners to engage in the work.

Second, although TMSJ embraces efforts to change the status quo, not everyone will be in agreement or accept that mathematics classrooms are the place to do this work. You may face pushback, hence the importance of using discretion within the community. Building a community of collaborators means building trust, which sometimes necessitates protecting both the group and TMSJ educators by being careful who you share your ideas with and assuring anonymity in public-facing materials. This protective stance allows everyone to concentrate on creating impactful experiences without undue influence from external forces that often hinder progress.

Third, it's crucial for TMSJ educators to unequivocally support each other in this endeavor. This means being unapologetic in this work, individually and collectively. You are in the position of helping children who will ultimately impact their respective communities; therefore, it is critical to remain steadfast and not waver in this work. You can demonstrate your unapologetic commitment by

- learning about each other;
- communicating regularly;
- visiting each other's classrooms virtually or in person;
- sharing lesson plans;
- brainstorming ways to address inequities in your classroom, school, and community;
- holding your colleagues and yourself accountable; and
- being available for feedback and support, especially if things don't go as anticipated.

Summary

In this chapter, you learned why building a community of trusted and dedicated collaborators is critical in the work of TMSJ. You learned the importance of having a core group of collaborators, how to identify collaborators, and strategies to grow and sustain a community of collaborators. Our hope is that you have been able to identify at least one person in your school or local setting that you can walk with as you continue your TMSJ journey. Now, it is time to start listing possible collaborators and the next steps.

To hear more from the authors on the topic of building a community of collaborators, listen to this **conversation with Dr. Childs and Dr. Staley.**

qrs.ly/lrffum2

REFLECT

Review your answers to Check In: In Search Of. Now consider how you "show up" to do the following:

- Support others in monitoring and checking their biases.

- Provide critical constructive feedback to peers on their classroom setting and interactions with children.

- Walk with others to learn about the children and school community.

- Design lessons that integrate school, community, and social justice issues into the classroom/school.

_____

_____

_____

_____

_____

_____

_____

_____

_____

_____

_____

 **ACT**

1. Add your key takeaways and next steps to your TMSJ Action Plan.

2. Someone to Walk Out My TMSJ Journey: Take some time to identify at least one person or group who you would like to collaborate with and include a brief rationale. Consider the pros and cons of inviting them into your space based on your understanding of TMSJ, your current needs, and what you are ultimately seeking to do as it relates to TMSJ experiences for your children. Use this as an opportunity to intentionally build your community. Schedule time to begin the conversation around the TMSJ work you are doing.

_____

_____

_____

_____

_____

_____

_____

_____

online resources ↖ A fillable template for Someone to Walk Out My TMSJ Journey is available for download at https://qrs.ly/wbfixtr

Where to Next?

As we near the end of this book, it is now time to start considering your legacy as an educator. Successful TMSJ experiences are limitless and endless. In the next chapter, we will explore the TMSJ educator mindset for crafting experiences that pass the test of time.

LEAVING YOUR LEGACY

As an educator, you never fully know the extent of your influence. Thus, it is imperative that we are intentional in the work that we do, as opposed to leaving things to chance and happenstance. It is helpful in this work to do some planning and consider the end of your career; take a moment and imagine the day after you retire. You are sitting on your porch in a rocking chair and reminiscing. What are the top three things you want to be known for as an educator, and why? Be big, bold, and unapologetic. Now consider what you must do between now and that day to achieve the three things. Just as you do with TMSJ, as you map out this trajectory for your career, the key is to be intentional and remain steadfast as you focus on that bigger dream. Now, let's go back and see how this dream aligns to TMSJ.

In this chapter, we will reflect on what we want from this career as an educator and how we will be remembered by

➤ understanding your why, and

➤ recognizing your influence.

Understanding Your Why

The work of a TMSJ educator is keenly focused on maximizing one's time and fulfilling one's purpose. Previous chapters have provided you with all the necessary information to become a fantastic TMSJ educator; now it is time to learn how the work continues beyond you. This section helps you uncover what drives you as a TMSJ educator, through the following questions:

➤ Why did you become an educator?

➤ So far, has your journey as an educator been fulfilling and joyful to you?

WHY DID YOU BECOME AN EDUCATOR?

Take a moment to reflect on the day you decided to become an educator. What thoughts and feelings arose on that day? Did you feel a sense of relief regarding your career choice? Did you feel a sense of anxiety?

Check In

In two to three sentences, write down your main reason or reasons for *becoming* an educator.

_____

_____

_____

_____

_____

_____

_____

Take a couple moments and reflect on what you wrote down, then write down why you are *still* an educator. Are these reasons the same or different from why you became an educator?

_____

_____

_____

_____

_____

_____

_____

_____

Now, let's take it a step further and reflect on your experiences over time. Jot your thoughts in response to the following questions:

Has a career as an educator been what you expected? If you could start your career over, what would you do the same? What would you do differently? What has been the biggest change you have noticed in education since you began your career? Finally, what is the biggest lesson you have learned as an educator?

_____

_____

_____

_____

_____

_____

_____

_____

_____

HAS YOUR EDUCATOR JOURNEY BEEN FULFILLING AND JOYFUL TO YOU?

Now that you have written down the biggest lesson you have learned as an educator, take a moment to reflect on what has been the *best part* of being an educator. What have you enjoyed the most? Why has this been the best part?

Now, switching gears, what is one thing you would like to have changed on your journey as an educator? Why would you like to change it? Is there anything you could have done differently? Take a moment to reflect deeply on how this one thing has impacted your career and in what manner. Record your responses in the space provided.

 **Check In**

What is one thing you would like to have changed on your journey as an educator? Why would you like to change it? Is there anything you could have done differently?

_____

_____

_____

_____

_____

_____

Now, let's focus on a different aspect of being an educator: joy. Every educator knows this can be a joyous and rewarding career choice. However, joy can often be a fleeting sensation, overshadowed or easily forgotten in the grind of daily work and life. It's essential to take a moment to lock it in. Over the course of your career as an educator, what have you found to be a continuous source of joy? Do not limit your thinking. Make this reflection as personal and deep as you can and write it down.

Check In

What are my continuous sources of joy?

_____

_____

_____

_____

_____

_____

Read aloud what you find joy in three times. Allow this to sink into your heart. Hold onto it as you continue your educational career. It will help you maintain your sense of dedication and balance as you think about the children under your purview and what this work will mean to them.

Recognizing Your Influence

Once you have had an opportunity to reflect upon your reasons for engaging in this work and crafted a vision of what you want to accomplish by the end of your career, it is time to think about the extent of your influence. This section answers the following questions:

➤ What does engaging children in TMSJ mean for your legacy as an educator?

➤ How would you like the children under your purview and peers to remember you?

ENGAGING CHILDREN IN TMSJ AND YOUR LEGACY

TMSJ is more than just a social justice math task in the classroom. It is an all-encompassing experience that transforms and impacts children's lives through the content, building their knowledge base, and having an impact on the community around them.

TMSJ is more than just a social justice math task in the classroom. It is an all-encompassing experience that transforms and impacts children's lives through the content, building their knowledge base, and having an impact on the community around them. Before stepping into TMSJ, it is important to look deeply into the mirror and ask yourself an important question: "Why do you want better for the children you teach?" As you begin to reflect on this question and your response, push yourself to move beyond a saviorism mentality and concentrate on how this work truly benefits the children under your purview. Will the children you teach be genuinely better off? This is a challenging question because it makes one really think about not only the work they are doing but the reason for doing it. It provides a foundation that can recenter you when the work becomes challenging.

During this recentering, it is helpful to revisit earlier chapters and why you felt it was important that children under your purview engage in TMSJ. Has your "why" expanded as you read through this book? If yes, how? If not, why not? TMSJ is a continual reflection experience, as one is constantly seeking to improve and become a better person and a better educator, with a keen focus on ensuring children fully benefit from the experience.

Just as we reflected on you as an educator and your ultimate outcome, let's now consider the children's ultimate outcome. What is one thing at the conclusion of a TMSJ lesson you want children to internalize? Why is this important to you? How will you ensure this happens? Taking it further, how will this align with your overall vision for your TMSJ experience?

BEING REMEMBERED

As you have learned, teaching mathematics for social justice is more than a pedagogical approach to teaching. It is a lifestyle. You don't just teach for social justice; you embody social justice. It is not just something you do; it is who you become. As an educator, you can impact children's lives in a multitude of ways, providing them tools and skills to succeed not only in a course but also in life. These tools and skills include coping strategies, coaching, developing a positive mindset, exposure to mathematics careers, opportunities to become civically engaged in their community, and applying a social justice lens to living life. These are all a part of the legacy you leave.

Sometimes it may feel as if the work you are doing in the classroom is insignificant; yet, it can have far-reaching impacts greater than you could have ever imagined. What happens in your classroom and the experiences your children take away have the potential to impact others in the school, the community, the region, and much farther, throughout the course of their lives. Having a TMSJ mindset helps you think about the future you are preparing your children for. As the old saying goes, "Children may forget what you taught them, but they will never forget how you made them feel."

> Having a TMSJ mindset helps you think about the future you are preparing your children for. As the old saying goes, "Children may forget what you taught them, but they will never forget how you made them feel."

➤ What feelings are you providing the children?

➤ What experiences are you providing them?

➤ What kind of effect will the learning experiences you are providing them have?

➤ How do your children's goals and dreams relate to your TMSJ efforts?

➤ How do you even know your children's goals and dreams?

 ## TRY THIS: MY CHILDREN'S DREAMS

Children need to dream big. To make sure you deeply know and understand them (as described in Chapter 3), and to inspire and motivate you to connect your instruction to their dreams, ask them to write down at least ten of their goals and dreams. That may seem like a large number, but you may be surprised how creative children can get about their futures when given the space and encouragement. Push them to dream super big. This will allow them to have something in mind that they are seeking to achieve and help change their outlook on mathematics and in life, and it will help you know them that much better and connect their biggest desires to your lessons.

To hear more from the authors about leaving your legacy, listen to this **conversation with Dr. Childs and Dr. Staley.**

qrs.ly/b3ffum5

Summary

In this chapter, you had the opportunity to deeply reflect on your why and your influence as a TMSJ educator. This chapter provided you time to truly consider why you are engaging in this work, how it will impact you, and how it will impact the children under your purview. In addition, you explored and began to take steps to secure your legacy as a TMSJ educator. Now it is time for your final reflection and action in this book.

REFLECT

Your journey as a TMSJ educator started the first time you taught a math lesson. Consider the following as you think about your why and your legacy.

- Revisit your response to Check in: My Vision (Chapter 1), make adjustments, and add insights to align it with the legacy you wish to leave.

- Add a brief statement to your vision to express your why. Why you do what you do. Your purpose, cause, or belief. Why you continue to press on in your TMSJ journey.

_____

_____

_____

_____

_____

_____

_____

_____

_____

_____

 ## ACT

1. Review your key takeaways and considerations for next steps from each chapter recorded in your TMSJ Action Plan. Prioritize them, noting those you have direct control over and those outside of your sphere of control. Identify the first action you will commit to taking and develop a plan with clearly articulated steps and a timeline. Now act!

2. Things You Love: Grab a stack of 3 × 5 index cards and record 180 things you love about being a social justice mathematics educator, one note per card. Then, every day of the school year, use these as your mantras and inspiration for engaging in the work.

_____

_____

_____

_____

_____

_____

_____

_____

Where to Next?

You made it to the end of the book. Congratulations! For some, this was a quick read; others may have read it slowly over time. Regardless, you have made it to the end, and now the real journey begins of planning your future as a TMSJ educator. Take a couple of days to deeply reflect on your career plans for the next 1, 3, 5, 10, and 20 years. Use the chart below to list some career goals related to teaching mathematics for social justice, your rationale, and the steps you will take to achieve the goals.

Planning for the Future

What are your goals and what is your plan?

Time Period	Goal(s)	Rationale	Steps to Take
1 year			
3 years			
5 years			
10 years			
20 years			

online resources Available for download at https://qrs.ly/wbfixtr

EPILOGUE

A Note From the Authors

From Dr. Kristopher J. Childs

Now that you have completed this book, understand this is not the end but just the beginning of your journey as a TMSJ educator. You are joining numerous other TMSJ educators who have read this book and are committing to moving the work of social justice forward. Understand that the work you will engage in from this day forward will live perpetually, and you are continuing the work of the many other TMSJ educators who have come before you. TMSJ is not a new phenomenon; we have just been privileged to enter this work at this moment in time and to make our contribution. To that end, we hope you will periodically take moments when engaging in the work to pay homage to those who have come before you, as it would not be possible without their hard labor and dedication, often in the face of fierce opposition. The trail has been blazed for all of us; thus, it's now up to you to move unapologetically forward as a TMSJ educator. You now have what you need to do amazing things in the work of TMSJ. Let no one or nothing stop you as you move forward. The children in your environment and future environments depend on you doing your part. You are like a domino in the middle of a row of dominoes. No one knows when the row of dominoes started or when it will end; however, we do know at the end is a truly equitable education system and society. The universe is waiting for you to fall forward and knock down the next domino, sending the combined energy and momentum forward. You play a critical role in advancing this work and making the handoff to the next generation. So, whatever you do after you close this book, move forward and Be Bold, Be Relentless, and most importantly, BE UNAPOLOGETIC!

—Kristopher

144

From Dr. John W. Staley

My career as a mathematics teacher and leader has been driven by issues related to equity, social justice, and the need to help students value and use mathematics as a tool to address social justice issues in their lives. My teaching career began in 1987 at a juvenile correctional facility for young boys between the ages of 12 and 18, where I realized that teaching mathematics meant more than teaching students the steps to find the "correct" answer(s) to problems. I quickly recognized that my calling to teach students extended beyond mathematics concepts and skills to lessons that they could use in this journey called life. From the beginning, my goal as an educator has been to develop my students' self-confidence and belief in their ability to try, do, and reflect as they live their lives, inside and outside of the mathematics classroom. Thus, I designed lessons and learning opportunities to encourage and empower students to become thinkers and doers of mathematics, connect to students' lives and hopes for the future, help students see relevance in the mathematics they are learning, and model for students the concept and value of respect—respect for self and others. I also realized that as a Black man from the city of Philadelphia, Pennsylvania, I had the opportunity and responsibility to be a positive role model and advocate for my students. This opportunity eventually extended to work with other educators and adults.

As I think back to where my journey started, I am reminded of a growing feeling that there had to be more I could do for the students in my classes; thus, the equity leg of my journey led me to envision a more meaningful, relevant, and fulfilling experience in the mathematics classroom for my students. My focus and understanding of equity in the mathematics classroom was initially developed by reading and reflecting on journal articles and books; engaging in conversations with many colleagues at conferences, workshops, and other in-person or virtual gatherings; and writing and speaking about equity and social justice. Next came a venture into CRP/T—culturally relevant pedagogy and culturally responsive teaching. Ladson-Billings's *The Dreamkeepers* (2009) and Gay's *Culturally Responsive Teaching: Theory, Research, and Practice* (2018) helped lay a foundation that I have continued to develop, in a similar manner as I did to build my understanding of equity in the mathematics classroom. And last, but not least, Teaching Mathematics for Social Justice (TMSJ). I remember the first time I picked up that blue and black book, *Teaching Mathematics for Social Justice: Conversations With Educators* (Wager & Stinson, 2012) and began trying to unpack TMSJ. Let's just say, it took many conversations with colleagues (thank you) and a few more readings by authors such as Aguirre, Berry, Gutstein, Gutiérrez, Martin, Peterson, Gorski, and TODOS's special issue "Mathematics Education: Through the Lens of Social Justice" (Aguirre & Civil, 2016) to help me make sense, deepen my understanding, and find my voice. I still, today, find myself looking back at charts like the Social Justice Mathematics Teaching Framework in this book (Chapter 1, Figure 1.1) that visually represent or summarize the connectedness and interdependencies of the "equity-related" pedagogies.

As I continue my TMSJ journey, reflecting on the many people and conversations, I am reminded to give myself grace and space as I continue to learn, realizing that there is still much work to be done. May you be blessed as you continue your journey.

—John

APPENDIX A

TMSJ Action Plan

 The TMSJ Action Plan is a tool that provides space for you to record your key takeaways from each of the chapters and next steps to support you as you continue through each stage of your TMSJ journey. The tool consists of three components:

Read & Reflect	At the end of each chapter, you will be prompted to take action: Add your key takeaways and next steps to your TMSJ Action Plan.
Review & Prioritize	In Chapter 10, we invite you to review your key takeaways and considerations for next steps from each chapter. Prioritize them, noting those you have direct control over and those outside of your sphere of control.
Plan, Do, Study, & Act	Plan, Do, Study, & Act (PDSA) is a continuous improvement cycle that provides a process for creating an action plan. Detailed steps are included in the PDSA template below.

Read & Reflect

Chapter	Takeaways	Next Steps
Chapter 1: Teaching Mathematics for Social Justice: What Is It and Why Does It Matter?		
Chapter 2: Mirrors: Understanding Yourself		

(Continued)

Chapter	Takeaways	Next Steps
Chapter 3: Windows: Understanding Your Children		
Chapter 4: Envisioning the Ideal Mathematical Environment for TMSJ		
Chapter 5: Engaging Children in the TMSJ Classroom		
Chapter 6: Planning a TMSJ Experience		

Chapter	Takeaways	Next Steps
Chapter 7: Implementing TMSJ in the Classroom		
Chapter 8: Taking Action Beyond the Classroom		
Chapter 9: Building a Community of Collaborators		
Chapter 10: Leaving Your Legacy		

Review & Prioritize

Review your next steps and identify two or three for possible action. Note those you have direct control over and those outside of your sphere of control. List them below and provide a brief rationale.

Next Steps for Action	Rationale

Plan, Do, Study, & Act

Select one of your actions from Review & Prioritize and develop at least one goal, then follow the steps in the PDSA cycle.

1. **Plan:** Identify needs and examine factors that lead to or perpetuate inequities. Note the questions you have about what you want to learn, make predictions, and identify evidence you will collect to test your predictions and monitor progress.

2. **Do:** Define the steps you will take to answer your Plan questions and a timeline for implementation. Implement your action and gather evidence of progress from multiple data sources.

3. **Study:** Analyze your data to determine your progress. Compare the results to your goals and report on your progress.

4. **Act:** Identify needed adjustments and determine your next steps. Then begin the next PDSA cycle.

Action: What action from Review & Prioritize will you take to continue growing as a TMSJ educator?

Goal: What is your goal (or goals) for this action?

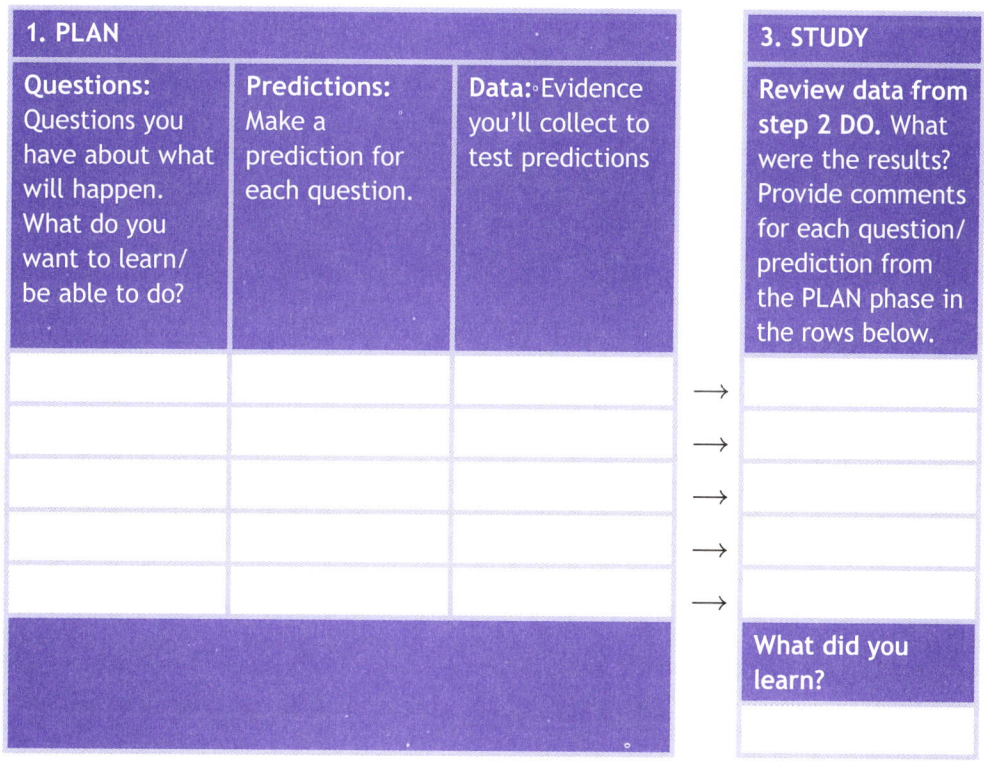

1. PLAN				3. STUDY
Questions: Questions you have about what will happen. What do you want to learn/ be able to do?	**Predictions:** Make a prediction for each question.	**Data:** Evidence you'll collect to test predictions		**Review data from step 2 DO.** What were the results? Provide comments for each question/ prediction from the PLAN phase in the rows below.
			→	
			→	
			→	
			→	
			→	
				What did you learn?

(Continued)

(Continued)

2. DO: What steps will you take? Set a timeline for each step.	**4. ACT:** What will you do next?

online resources — Available for download at https://qrs.ly/wbfixtr

APPENDIX B

Family Letter About Math Class

Dear Families,

Welcome to _____ mathematics classroom. Throughout the year, we have focused on developing the children's problem-solving skills. I have been intentional in ensuring the children are aware of the real-life applications of mathematics. In class, I am challenging them to think deeply about the material, and they are given opportunities to showcase their understanding. As the children explore real-life applications, they are seeing how to use mathematics to improve society in a variety of facets. Teaching mathematics for social justice focuses on engaging children in mathematics learning experiences rooted in social issues, with an emphasis on transforming their communities.

In the upcoming units, we will focus on four pillars:

1. The use of rich problem-solving tasks

2. The children identifying a social issue or issues

3. The children engaging in mathematics rooted in a social issue or issues

4. Mathematics being used to transform their communities

Your child will use these experiences to strengthen their understanding of mathematics and make real-life connections. Through these experiences, your child will be better prepared for their next mathematics class and for future mathematics experiences. As we engage in these experiences, I will continue to communicate with you regarding the social topics they deem of interest, as these experiences are very collaborative. As always, I appreciate each of you and your continued support of the mathematics experiences the children are receiving. If you have any questions or concerns, please do not hesitate to reach out.

Mathematically yours,
Your Child's Teacher

APPENDIX C
SAMPLE TMSJ LESSONS

We would like to take a moment and thank the authorship teams of *Mathematics Lessons to Explore, Understand, and Respond to Social Injustice* series (2020-2022) for your work to help educators at all grade levels. To the many educators who provided lessons for each of the books, we thank you for sharing a part of you, as many of the lessons come from your school, community, background, family, and lived experiences.

A special thank you to the following authors for your lessons that are included in this book.

Note: These sample lessons are directly from the series and reflect the lesson numbering from the original publications.

Koestler, et al. (2022). *Early Elementary Mathematics Lessons to Explore, Understand, and Respond to Social Injustice.* Corwin.

- Lesson 5.4 Examining Air Quality by Maria del Rosario Zavala
- Lesson 5.13 Early Elementary Mathematics to Explore People Represented in Our World and Community by Courtney Koestler, Eva Thanheiser, Mary Candace Raygoza, Jeff Craig, and Lynette Guzmán

Bartell, et al. (2022). *Upper Elementary Mathematics Lessons to Explore, Understand, and Respond to Social Injustice.* Corwin.

- Lesson 5.1 Families Matter by Nicky Meindl
- Lesson 5.6 Challenging Ableist Assumptions in Mathematics Problems by Courtney Koestler, Jennifer R. Newton, and Jan McGarry

Conway, et al. (2022). *Middle School Mathematics Lessons to Explore, Understand, and Respond to Social Injustice.* Corwin.

- Lesson 7.2 The True Cost of That $29 T-Shirt in the Store Window by Bethany Chan, Debasmita Basu, Rebecca Ellis, Frances K. Harper, and Jennifer Ruef
- Lesson 9.4 How Many Meals Can Minimum Wage Buy? by Elizabeth O. Ayisi and Colleen Carman

Berry, et al. (2020). *High School Mathematics Lessons to Explore, Understand, and Respond to Social Injustice.* Corwin.

- Lesson 5.3 Listen to GLSEN by Bryan Meyer and John W. Staley
- Lesson 7.5 Humanizing the Immigration Debate by Ayensur Ozturk and Steve Lewis

LESSON 5.4 EXAMINING AIR QUALITY

Maria del Rosario Zavala

Air Quality

In November of 2019, schools all over California were learning to deal with a new and dangerous season: smoke season. While California is no stranger to wildfires, the amount and size of the fires that arose in the 2019 fall season, when California is at its driest, created the new phenomenon of air so unhealthy that health experts started issuing guidelines for children to stay inside. "Spare the air" days, which are days where the air quality is so unhealthy that it's recommended to not engage in strenuous exercise, gave way to "stay inside" days. Most San Francisco Bay Area schools were built for the mild climate that was the norm in the mid-20th century: large outdoor play areas, classrooms that are connected by exterior hallways, and, perhaps most problematic, no central air systems and many older or portable buildings with no air conditioning.

When the first unhealthy air day was announced, children suddenly were inside during recess. Teachers were asked to keep windows and doors closed to limit poor-quality air coming inside, which only led to further issues. Understandably, kids were looking at the window and seeing no rain, and wondering why they needed to stay inside for recess on sunny days. This lesson is designed to help children explore what air quality measurements systems are, and how they relate to whether air is safe to breathe or not. While this lesson does not go extensively into the science of measuring air quality, it could be a springboard into those kinds of questions.

SOCIAL JUSTICE OUTCOMES

- I see that the way my family and I do things is both the same as and different from how other people do things, and I am interested in both. (Identity 3)

- I know that life is easier for some people and harder for others and the reasons for that are not always fair (Justice 14)

MATHEMATICS DOMAINS AND PRACTICES

- Number and Operations

- Measurement

- Data Collection and Analysis

- Make sense of problems and persevere in solving them. (MP1)

- Construct viable arguments and critique the reasoning of others. (MP3)

- Model with mathematics. (MP4)

CROSS-CURRICULAR CONNECTIONS

- Language Arts

- Science

- Social Studies

Deep and Rich Mathematics

The mathematics in this lesson are meant to support number concepts central to kindergarten. Many children in kindergarten are learning numerals, number names, and rote counting by ones. They are making sense of quantities and how our number system is ordered, exploring ideas like how the more digits a whole number has, the bigger it is. Alongside this, they are reasoning about the meaning of amounts and are learning concepts to compare numbers such as "more than" and "fewer than."

More than the particular concepts of number, a central goal of the lesson is to help the children connect numbers to the world outside. The very idea that we can and *do* measure things like particles in the air through some kind of scale is a mathematical goal in this lesson. This lesson is intended to help plant a seed to continue talking about air quality at home, to have some sense of why it's measured, and to show how the number that is measured impacts how we choose to act.

About the Lesson

The lesson is introduced in 1 day, but ideas can be revisited in subsequent days. The lesson intentionally has variation in structures for participation: whole group, small group, and then whole group again. Depending on what kinds of resources you have (e.g., a projector), you may need to gather printed-out materials the day prior to the lesson, including large-format images of different regions of the world with varying air quality. You should make the materials easily accessible for all children at their table groups. (Examples of maps are listed in the *Resources and Materials*.)

This lesson can be done in 1 day, over an approximately 45-minute chunk of time. While this lesson highlights specific locations near the school, teachers could modify these locations to include those relevant to the children in their particular context.

Although this lesson was designed for and taught in a kindergarten classroom, it would be appropriate for preschool and primary classrooms. Additional adaptions could be made for other grades, including higher education methods coursework. One suggested extension for older children is included after the lesson flow.

RESOURCES AND MATERIALS

For the lesson, you need the following:

- Maps of relevant locations with varying air quality. One good source is PurpleAir (https://www2.purpleair.com), especially on days when local areas have poor air quality. Therefore, you may need to plan ahead and collect images of maps from days that are of interest because of the range of air quality. Otherwise, you can search AirNow's archival data and see if you can find relevant map data (https://www.airnow.gov/).

- Teacher Resource 1: *Maps for Exploring Air Quality*. This resource provides an example of the different ways two sites show data and the maps described in the lesson.

    + The images utilized when this lesson was initially designed, and therefore recommended as a starting point, are as follows:

        - Lake Tahoe, California, because many children go there with families in summer and winter, with really good air (green dots);

        - New Delhi, India, which is the city with the distinction of the worst air quality on average (dark purple dots);

        - The entire state of California, which has a variety of dot colors, given the fires; and

        - The school with a two-block radius to include a few sensors, with a variety of dots ranging from green to dark orange.

The following may be helpful too:

- A large image of an air quality table (see Teacher Resource 2: *Three Ways to Display How Air Quality Is Measured*)

- Air quality "number line" meter

- Small stickers for children to predict air quality on the meter

- Teacher Resource 3: *Sample Family Letter on Supporting Children to Keep Exploring Air Quality*

Prior to teaching the lesson, here are suggested resources to develop background knowledge on air quality for yourself.

- Article: "Study finds wildfire smoke more harmful to humans than pollution from cars," NPR (https://n.pr/3d6Gr5c)

- Article: "Long wildfire seasons also mean extended periods of dangerous air quality," NPR (https://n.pr/3rqxax0)

- Radio story: "Smoky air from wildfires impacting parts of California differently," KQED (https://bit.ly/3p-dt7S6) [Smoke story starts at 5 minutes]

- Website: California Air Resource Board, "Children's Environmental Health Protection Program" (https://bit.ly/3odWpAX)

Lesson Facilitation

Launch (5 minutes)

Introduction to the Maps

Call children over to sit on the rug in spots or in rows. Show a map of California (or the state in which the class is located). The goal of the launch is to establish that the class is looking at maps that show them air quality readings.

- Show the map of the whole state of California, and ask children if they know what this is. Give children some time to observe the map and point out what they notice about it. Ask questions to facilitate exploration, and ask children to listen and respond to what other children say. Support them in making connections to the numbers and colors on the map.

+ Ask questions such as these:

+ *What does the map show?*

+ *What colors do you see?*

+ *What numbers do you see?*

+ *What do you think different numbers and colors mean?*

+ *Is this a big number?*

+ *And it means what?*

+ *Why are we reading this map?*

+ *What does it help us see?*

+ *Could we just go by the color, or does knowing the number help too?*

• Elicit connections between the map and air quality through a brief discussion about air quality, and why we would measure it.

+ Ask: *What does air quality tell us? How have we been affected by air quality lately?*

• Next, have children transition to their seats to explore the maps in small groups.

EXPLORE (APPROXIMATELY 10 MINUTES)

Children Explore the Maps in Small Groups at Their Tables
Let students know that each table group is going to get a map, and one table group will get the map you have all just been looking at. Directions can be given while children are still in the whole group.

• One group should get the map being viewed in the launch, another group should receive a map from their local neighborhood, one of New Delhi, and other locations as determined by the local school context. Some of these maps should include locations children are familiar with (in this case, Lake Tahoe).

• At their table groups, children will repeat what they did in the intro-duction: look at the map, and take turns noticing and wondering.

+ *I see . . .*

+ *I know that . . .*

+ *I wonder if . . .*

Tell them you will circulate and listen to what they are noticing and wondering. It's important to let go here and listen to what children are noticing. That is the priority right now, to privilege students' observations and validate their contributions.

- You can circulate and help make connections to their maps. For example, say: *This is a map of Lake Tahoe. Have any of you been to Lake Tahoe? What do you go there to do? I see green dots with numbers around the lake. What do you think that means?*

- Ask follow-up questions to what students are noticing. Help students read aloud large numbers, if students are not sure how to say them. For example, ask: *What do you think different numbers and colors mean? Is this a big number? What is it counting? I'll read this number for us just in case: three hundred twenty-four. Does that sound like a high or a low number?*

- Prompt students to share different ideas when you return to the rug. It might sound like this: *Can I ask you to share that idea, Simone? You just said, "There's only small numbers like 5 and 6 on our map." Will you share that when we get back to the rug?*

CONNECT (15 MINUTES)

Children Make Comparisons Across Maps

Have children come back to the rug, this time in a circle. Put all maps in the middle of the circle, along with an air quality table.

- This time should be carefully structured so children are listening to each other. You might also prompt children to look at and speak to each other, not just you as the teacher. This helps facilitate how the conversation is happening between all of us, not just individual child to teacher.

- Prompt a representative from each group to share a key thing their group noticed, one by one. After each group shares, contrast the two maps with the most extreme measures (in this case, Lake Tahoe and New Delhi; the Lake Tahoe example may change based on the location of context of the lesson).

- Ask children if they think these two places have very different air qualities. For example, ask: *Which place would have been safer to be outside without a special mask for breathing? How do we know?*

From here, support children to make sense of the map of their neighborhood. Hold up the map of the neighborhood around the school and revoice the observations of the group that had this map.

- Facilitate a discussion about how different families are managing with changing air quality. Ask: *Did you go outside yesterday? Did you take precautions?* Students can share their experiences, listen to each other's activities, and notice similarities and differences in how families adjust for changing air quality.

SUMMARIZE (5-10 MINUTES)

Students Take a Position and Support It With Evidence

Tell children that they can use all of this new information to make a decision about whether they can go outside. The teacher holds up the air quality map of the area surrounding the school. In this case, the monitors are giving us different readings, but the principal of the school has to make a decision. The goal of this time is to connect it to the local context. If the decision were up to children, using the information we have, would they declare this an indoor or outdoor recess day?

- Ask children to think, *Do you think the air is mostly good, mostly moderate, or mostly unhealthy? Should we stay inside for recess? Why or why not?*

  + Children may not use the data presented to draw their conclusions. Be ready to prompt them to think about the color of the dots on the map and the size of the numbers.

- Ask for several volunteers to place stickers on the "air quality number line" where they think today is best represented. Ask them to share why they placed it there and why.

  + Once a few children have shared ideas, you can close the lesson by reminding them to listen over the next few days for when people are talking about air quality on the news and with their families, and think about the numbers behind the colors.

  + Alternatively, you can ask pairs of students to discuss and decide. The goal here is not for children to decide whether they are right or wrong, but rather to use ideas of air quality in how they decide. You should be ready to listen carefully and prompt children to make connections back to what the map around our school tells us about air quality.

- Share with children: *When you are outside and smell smoke, or see a lot of gray haze, you can ask the adults in your lives to help you look up the air quality to make decisions about going outside.*

You can play the 5-minute KQED radio story on the different impacts of smoke across California (https://bit.ly/3pdt7S6; the smoke story starts at 5 minutes). Ask the children to write down any numbers they hear, noting the following:

+ *What number did you hear?*

+ *What did it mean?* (for example, *I heard 1 month. She said that's the amount of smoke days people experience in a year now.*)

+ *Why would they include that number in the news story?*

This activity is another way to help students connect mathematics to the world around them and to see how numbers and quantities are utilized to communicate information, such as information about the air we breathe.

TAKING ACTION

Suggestions

Individual and Class

The suggested action would be for students to talk about air quality with their parents, to listen at home for when air quality was being talked about on the news, and to have more context for when recess is indoors. The lesson can also serve as part of ongoing conversations about air quality in relation to environmental justice, which could be a thematic unit the children study as some part of the school year.

Local Community, Organizers, and Organizations

Work with local agencies to explore what community/local efforts are happening to study and improve air quality. Share these ideas with school and home communities. Support the school, children, and families in partnering with these agencies.

Communicating With Stakeholders

Families

Send home a note to parents/caregivers on what the class had talked about and encourage their conversations with their children about the impacts of air quality on our health and why it's important to pay attention to air quality. See the suggested text in the *Resources and Materials* for this lesson.

- Either before or after the lesson, you can reach out to classroom parents and request that they ask their children what they did with air quality maps at school that day. Parents can ask children what they noticed on the maps, what other people said, and then if they have any questions. In this way, parents are supporting the students to notice, wonder, and ask and answer questions. They are cultivating their students' curiosity as well as supporting key mathematics and science practices about asking questions and engaging in problem solving. An example letter is provided in the online resources (Teacher Resource 3).

OTHER SCHOOL/DISTRICT PERSONNEL

- Create posters to put around the school about why recess is sometimes indoors. Younger children might pair with older children to make them.

- Create an air quality monitor and incorporate it into daily calendar routine for a month or so following the lesson. Teachers can make a bar graph and retrieve the measurement from nearby monitors on PurpleAir (https://www2.purpleair.com), either on a laptop or their phones, and children could fill in the bar graph for the right range. It may be sufficient for the teacher to include just the first four categories and make the fourth one a catch-all for any reading over 151. It should also be anticipated that we have all good or mostly good days, with some moderate days, which would still have mathematical value.

Online Resources

online resources ▸ Available for download at **https://qrs.ly/wbfixtr**

▼ Teacher Resource 1: Maps for Exploring Air Quality

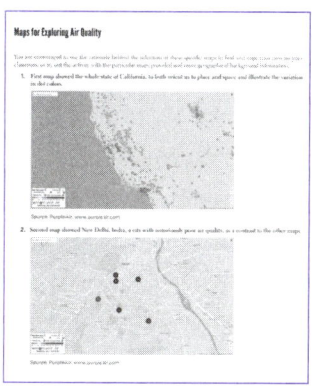

▼ Teacher Resource 2: Three Ways to Display How Air Quality Is Measured

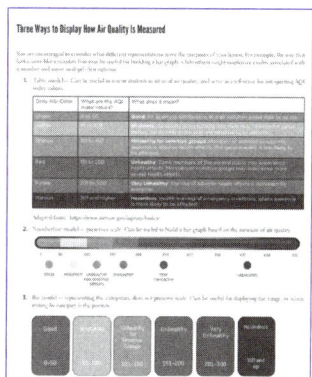

▼ Teacher Resource 3: Sample Family Letter on Supporting Children to Keep Exploring Air Quality

Background of the Lesson

As residents of California will tell you, we have had "spare the air" days as long as we can remember—days when the state's air quality board would send out announcements to our radio stations to tell people not to exercise too strenuously or burn unnecessary fires if they weren't needed for heat or cooking. Spare the air days are a result of taking a scientific approach to the smog and smoke that are part of living in our state. However, even as clean-air vehicle initiatives have been successful and industrial pollution has declined in California, we have seen a rise in wildfires. This rise has led to an increase in poor air quality days, and fears for children's health should they breathe the cancerous particulates released during wildfires, especially fires that burn toxic materials used in housing. This is also to say nothing of the families who lose their homes to wildfires, and the trauma left in the wake. In my community in Oakland, California, a community that remembers a devastating fire in the Oakland hills in 1991, children are growing up with multiple days out of the year in which they are kept inside or kept home from school due to poor air quality. This is exacerbated by the COVID-19 pandemic, as protocols require children to eat lunch outside because there is less possibility of COVID transmission, but there is still the possibility of poor air quality.

When this lesson was developed, it was prior to the start of COVID, but not much. My son was a kindergartener and I was volunteering in his class. There were 2 days in a row that children were not allowed outside for recess, and it was frustrating for both the adults and kids. I asked the teacher if I might do a lesson on air quality to help children understand that even if they couldn't *see* why they couldn't go outside, there was a reason. The air quality lesson was developed as a result of me virtually spanning the globe through the PurpleAir.com map and looking for places where we could contrast the air quality. With the visual anchors of the maps, the rest of the lesson fell into place.

When I did this lesson with the kindergarteners, it was November of the school year. The lesson was launched in a whole-group setting with noticing and wondering about the map of California. After a few minutes of sharing what they noticed and wondered and establishing that this was a map of the state of California and that it had different-colored dots, the children were dismissed to their table groups to pore over one of the maps and prepare to report back on what they saw on their map. What color dots were there? What numbers? What do they think the dots and numbers mean? Most students thought the numbers were street addresses.

After a few minutes in small groups, they returned to the rug. A representative from each group shared their noticings. The table for air quality and concept of air quality was explicitly introduced at that point. We focused on the contrast between the Lake Tahoe map (great air quality, low number also indicated by green) and our school map (not great air quality, in the low hundreds, indicating harmful for sensitive groups). We spent the last few minutes trying to decide if the map of our area meant we should or should not go out for recess. Finally, students were reminded to talk with their parents about air quality, listen to the news for when air quality was talked about, and know that there are numbers behind how we decide that air is good or bad to breathe.

About the Author

 **Maria del Rosario Zavala, PhD,** is an associate professor of elementary education at San Francisco State University, with a focus on culturally responsive mathematics teaching, mathematics identity development, and bilingual education. She had various roles in education spanning K–12 schooling prior to her role at SFSU, and she continues to work in classrooms whenever she has the chance. In terms of social justice, she gives credit to her college professor at the University of California, Santa Cruz, Dr. Julia Aguirre, whose class she took as an undergraduate mathematics major. This course helped her to call out and start to really wonder about inequities in mathematics education and to make sense of her own experiences as a bilingual Latina woman. She views the teaching of mathematics as open to creativity, with limitless opportunity to connect to issues that impact children's communities and lives.

- I can describe some ways that I am similar to and different from people who share my identities and those who have other identities. (Diversity 7)

- I want to know about other people and how our lives and experiences are the same and different. (Diversity 8)

- I find it interesting that groups of people believe different things and live their daily lives in different ways. (Diversity 10)

- I know that life is easier for some people and harder for others and the reasons for that are not always fair. (Justice 14)

MATHEMATICS DOMAINS AND PRACTICES

- Number and Operations

- Data Collection and Analysis

- Make sense of problems and persevere in solving them. (MP1)

- Construct viable arguments and critique the reasoning of others. (MP3)

- Model with mathematics. (MP4)

- Attend to precision. (MP6)

CROSS-CURRICULAR CONNECTIONS

- Language Arts

LESSON 5.13 EARLY ELEMENTARY MATHEMATICS TO EXPLORE PEOPLE REPRESENTED IN OUR WORLD AND COMMUNITY

Courtney Koestler, Eva Thanheiser, Mary Candace Raygoza, Jeff Craig, and Lynette Guzmán

The Diversity of the "Global Village"

The social justice topic that we explore in this lesson is the diversity of our world and how it is represented mathematically. Children explore a global context through a lens of critical literacy where they analyze a picture book as a non-neutral text, with the idea that "all texts are created from a particular perspective with the intention of conveying particular messages" (Vasquez, 2016, p. 3).

By shrinking the world into a village of 100 people and learning about different aspects of their lives—nationalities, languages spoken, religions practiced (or not!), access to drinking water, those living in poverty, etc.—"we can find out more about our neighbors in the real world and the problems our planet may face in the future" (Smith, 2011, p. 7). Exploring a global context through this lens can (1) encourage children's critical inquiry about all of humanity's access to basic human needs and distribution of resources, (2) recirculate stories about ourselves and our neighbors, and (3) open up conversations to learn from and with each other. Furthermore, engaging with these data can be affirming and/or challenge assumptions about who lives in our world and facets of their lives, empowering everyone to be more aware of and connected to the global community. Together, we resist U.S. individualism and U.S.-centric curriculum and stories about humanity.

Deep and Rich Mathematics

This lesson prioritizes mathematical ways of comprehending problems. This includes starting problems using mathematics and mathematical language, making sense of mathematical relationships through decontextualizing and contextualizing problems, and utilizing mathematics to reconsider perspective. To accomplish these goals, children should be constantly reminding themselves that the World as 100 People is a lens for framing any problems they identify and analyze. Children use various representations (concrete such as Unifix cubes, pictorial representation, number lines, expressions, and graphs) and compare across them. Children also learn how to find data and then represent them.

About the Lesson

This lesson highlights the value of visual representation as it helps to comprehend and analyze problems. Children can use many forms of representing these data, and they might see value in one representation over another through comparison and contrast. Children should be encouraged to view data visualizations as another conjugate within mathematical language, interconnected with the symbolic and numeric. Discussions should explore how certain kinds of representations make visible certain features of the given contextual situation, which might lend itself to particular inferences and meaning making.

Children can also begin to unpack the significant and nuanced discussion about objectivity and subjectivity, as it relates to mathematics and statistics. You should introduce or reintroduce the concepts of counting, sampling, and bias and demonstrate appropriate ways to question data without undermining the complexity of the task. In particular, you should assist children to approach data both without dismissive skepticism akin to these data being made up and without an undue idealism about the origins of these data. Children should discuss complex statistical concepts like precision and accuracy, alongside bias and subjectivity.

We anticipate this lesson to take between 2 and 2.5 hours in total, which could be split up across a series of days as needed. We include connections to mathematics domains that branch multiple grade levels, which could be modified based on the primary grade of focus.

RESOURCES AND MATERIALS

- Book: *If the World Were a Village: A Book About the World's People* by David J. Smith (2011)

- Website: 100 People, "A World Portrait" (https://www.100people.org/statistics-100-people/)

- Video: "The 100 People Project: An Introduction," from 100 People (https://www.100people.org/the-100-people-project-an-introduction/)

- **Note:** On the 100 People website, "gender" is the first item discussed but it is reported as a binary using markers of male and female. This is problematic because it doesn't unpack the diversity of gender (or sex). This is a great opportunity for critical literacy. (The same issue happens in most videos related to this topic.) Thus, one suggestion is to start with the book by Smith and then follow up with the 100 People video.

- Poster paper

- Markers

- Rulers

- 10 × 10 grids

- Templates for bar graphs, pie charts cut into 100 pieces, and so forth

- Concrete materials

- Local data about race and ethnicity, languages spoken, and other topics children may be interested in (school-level, city-level, and state-level data can be appropriate)

- Teacher Resource 1: *Supplementary Video Links With Social Characteristics in Each*

Lesson Facilitation

Day 1
GREETING (10 MINUTES)

- Begin the lesson with an opening circle greeting and community building activity entitled, *I love my neighbor who . . .* or *I have solidarity with people who . . .* The purpose of this activity is (1) to support children to get to know one another better and to build trust and community and (2) to challenge assumptions about one another and identify similarities and differences with one another.

- Have children circle up, seated in chairs, except for you in the middle.

- **Note:** It is suggested that you start in the middle of the circle to model the first prompt; you should participate fully as well.

- Begin by saying:

  + *I love my neighbor who . . .* and fill in the sentence with a trait or experience that is true for a child in the circle (e.g. "I love my neighbor who speaks Spanish").

- For whoever else it applies to, they must stand up and move to a new chair. The person who doesn't find a chair is the new person in the middle, and the process is repeated, similar to the game *Musical Chairs*.

- Prior to engaging in the activity, children should revisit classroom norms together (or create new norms) to ensure everyone is respectful and also knows they can choose not to participate or pass on particular prompts. In addition, children can brainstorm possible topics to include so that they are not caught off-guard thinking on the spot and so that a range of topics come up (e.g., birthday month, hair color or texture, favorite music genre, favorite hobbies). After children suggest topics, you may also offer some suggestions.

- **Note:** The activity will need to be designed around/accommodate for everyone's physical abilities.

- Close with a reflection and invite children to share what they learned, perhaps with one of the following sentence starters:

    + *I appreciated learning . . . [about one another].*

    + *We should not make assumptions about people because . . .*

LAUNCH (15 MINUTES)

- Introduce the idea of shrinking the world's population down to a village of 100 people that represents it by reading the first few pages of *If the World Were a Village* by David J. Smith. Use a globe or map to provide context if necessary. (Supplementary websites or videos can also be used; see Teacher Resource 1.)

- After reading the introduction, ask:

    + *In the village of 100 people, how many people do you think would live in the United States [or North America]?*

    + *How many people do you think would speak English?*

- You may also engage children in wondering about or predicting other characteristics of the world's village discussed in the book (for example, the ages of the residents of the village, how many residents have regular access to electricity, or other aspects that you think the children would find interesting).

- Continue reading the book to explore the characteristics of the village. (We typically focus on reading the pages about Nationalities, Languages, and Religions because of the way that the data are presented, but you can also read other pages if children are interested.) You can also explore the video "100 People Project: An Introduction" (https://www.100people.org/the-100-people-project-an-introduction/).

- Next are examples of how you might support children in exploring the data, while at the same time unpacking the text using a critical lens.

- Review the Nationalities data with the children. Ask:

    + *What do you notice about the data?*

    + *What do you wonder about this data?*

    + *How do you feel about this data?*

- Children will typically notice, wonder, or ask the following. You can also bring up these ideas, making sure to provide a lot of space for children's ideas first.

  + *Five villagers are from the United States and Canada, but later on the page it says that 5 are from the United States. What do you think this means?*

  + *The author states that there are 9 villagers "from South America, Central America (including Mexico), and the Caribbean" (Smith, 2011, p. 8). Why do you think the author decided to put Mexico with Central America when other continents are listed?*

- You might also want to explore follow-up questions:

  + *What exactly does nationality mean?*

  + *Who decides one's nationality?*

  + *How are language and culture related to nationality?*

  + *How many continents are there?*

- **Note:** Wikipedia has an interesting and informative page about continents (https://en.wikipedia.org/wiki/Continent), including a section about different ways of distinguishing the number of continents. You can discuss the socially constructed and contested nature of this information.

- Also ask:

  + *How could we represent this data?*

- Brainstorm with children multiple ways of representing data (e.g., conventional ways such as bar charts, pie charts, and pictograms) and unconventional ways such as hundreds charts (10 × 10 grids) using Unifix cubes.

- Take some time to brainstorm with children. Describe how they can use blank poster paper or templates of graphic organizers (bar charts, pie charts, 10 × 10 grids, Unifix cubes) to represent the data.

- While children are discussing representations, you should also ask them how different representations might communicate the data in different ways. For example, a pie chart shows the data in relationship to each other, whereas a bar graph more distinctly shows each piece of data.

EXPLORE (45 MINUTES)

- Assign groups of four children to explore various topics. (Typically, the Nationalities and Religions data work well to begin with because of how the data are represented. The Languages data are also typically an interesting data set, but also introduce some complexity.)

- Allow for some independent think time to explore the topic and have each child sketch various representations for the topic. Then allow small groups to share their ideas and to decide on several representations to include on a shared poster. On the poster, they can freehand draw their representations or use templates to create their representations, depending on their experience with data representation. They should be encouraged to include titles, labels, and keys as well as try to show the connections between and among the different representations.

SUMMARIZE (15 MINUTES)

In an effort to use more inclusive language, we use the term *gallery review* for what is commonly referred to as a "gallery walk."

- After all posters are created, groups can display their posters and the class can do a gallery review to observe. Gallery reviews can be done in different ways; for example, children can do the gallery review silently, simply looking at the posters up close or discussing them in small groups. They could also have sticky notes where they are able to pose questions to the group, which the group members later answer during presentations. After the brief gallery review is complete, children can share in various ways about what they learned about the data and how their representations communicate these data.

- Tell children they will continue the work the following day.

Day 2

Note: For Day 2, you will need to find data.

LAUNCH (20 MINUTES)

- Revisit the representations children made on Day 1.

- Use prompts such as these:

  + *What do you notice about the representations you made? What do you wonder?*

  + *What is still confusing?*

  + *What is surprising about the data?*

  + *What questions do you have about the data or the issues related to the data?*

> + *What did you learn about our world from this data?*
>
> + *What did you appreciate working with one another here in our classroom community?*
>
> + *What do you want to know more about?*

- The two following questions are especially important to get at the critical literacy part of the lesson:

> + *How do your **different** representations communicate the data **differently**?*
>
> + *What assumptions went into the presentation of the data in the book?*

EXPLORE (45 MINUTES)

- Find data about a context that you and your children are interested in—the country, their state, their city or town, and their school districts or schools. Public school and district data are also often easy to find on district or state board of education websites. (David J. Smith has also written a book called *If America Were a Village*.)

- Ask children what they notice about the data. Is the diversity (e.g., in terms of race and ethnicity, languages spoken) they see in the data represented in their own classroom? In their grade level? Why or why not?

- Repeat the process from Day 1, exploring data from children's city, town, school district, or school. (Have children create representations and discuss how different representations communicate the data in different ways.) This part could either immediately follow Day 1 or take place on another day.

- Then use these representations of the local data to make comparisons to the world data. Ask children what they notice and wonder about these comparisons.

TAKING ACTION

Individual, Class, or School

- After the lesson, children can create infographics to articulate what they have learned to their parents and caregivers about the diversity of the world to embody the Learning for Justice Standards (Diversity 8 and 10). They can use these infographics to compare their local communities to the greater world community (Diversity 7).

- Once children explore the diversity of their local context, they often are curious about resources and supports for different peoples. Teachers can engage children in using knowledge they gain to take action. For example, when children in one second- and third-grade classroom found out that there were a large number of children and families who spoke Spanish and Chinese but their school only had a Spanish Family Liaison support staff member working at the school, they wrote letters to the principal to find out why and to ask if there was any way to get funding for the Chinese-speaking families in the district.

Local Community, Organizers, or Organizations State, National, or Systems Level

- You may also support children to identify people (e.g., family members, politicians) or organizations (e.g., their own school) that may hold assumptions about who is represented in different spaces and to think about calling in or calling out those assumptions with data.

Communicating With Stakeholders

- Before teaching the lesson, you can let children's families know that you will be exploring the diversity in our world, both at the global and local levels. You can invite family members in to share their strengths, resources, and insights with the children. For example, if there are multilingual family members that would be willing to come in and share (e.g., by reading a book in a language other than English), invite them.

Online Resource

 Available for download at **https://qrs.ly/wbfixtr**

Link	Where from	Languages	Gender	Age	Religion	Shelter	Food	Water	Toilet	Literacy	Education	Income/Wealth	Electricity	Cell Phone	Internet	Health
If the World Were a Village of 100 People: https://youtu.be/PtYjUv2x65g	X	X														
The 100 People Project: An Introduction: https://www.100people.org/the-100-people-project-an-introduction/ https://www.100people.org/wp/the-100-people-project-an-introduction/	X		X		X			X		X	X			X		
If the World Were a Village of 100 People (2019 Edition): https://youtu.be/aLjIFo0TJfY	X			X	X	X	X	X			X	X	X			
If The World Was 100 People (Jay Shetty): https://youtu.be/LXqOd5noN8g	X	X	X		X			X			X	X	X	X	X	
If The World Were 100 People (GOOD Data): https://youtu.be/QFrqTFRy-LU	X	X	X	X	X			X			X	X	X	X	X	X
What If Only 100 People Existed on Earth? https://youtu.be/UbffuGZHeRO	X	X	X	X	X	X	X	X		X	X	X		X	X	X
If the World Was Only 100 People (Knovva): https://youtu.be/A3nEBT9ACg	X	X	X	X	X	X	X	X	X	X	X			X	X	X

▲
Teacher Resource 1: *Supplementary Video Links With Social Characteristics in Each*

Background of the Lesson

This lesson was developed to allow children to learn mathematics while making meaning about the world. We have all used this task in many venues, including in K-12 mathematics classrooms, in professional development settings, and with parents and caregivers. Each time we implement it, we all learn something new about our world. Eva first learned about this task from Courtney and adapted it into her own courses (and Courtney learned it from their friend and colleague, Ryan Flessner). After realizing we had each engaged in developing mathematical tasks that were similar yet distinct in focus and approach, we decided to come together to create this lesson for the book. Eva notices that children in her classroom are always impressed that they learned new things both about the world (e.g., how relatively few people of the 100 are from the United States as compared to their estimations) and about mathematics (e.g., that percent relates to per 100). She has been modifying this task over many years to then connect various things about the world (e.g., learning about the world and comparing it to a local city, country, etc.) and about mathematics (e.g., comparing fractions when comparing the world to a local city, country). Mary incorporated a version of the task in the first week of Algebra I, to offer children an opportunity to see how mathematics reveals important information about the world, sometimes challenging our previously held beliefs and telling a powerful story about people. Furthermore, it supported children to see the importance of working with fractions, decimals, and percents before launching into algebra. Lynette and Jeff helped develop parts of this lesson together, when Jeff was designing a course in quantitative literacy. Although that course was for undergraduates, we found the spirit of the lesson to be flexible, so Lynette adapted parts of it for her preservice elementary mathematics teachers to engage and unpack. We are always encouraged by the lesson, both in terms of the mathematical learning it can support and of the impactful ways children engage with questions of justice. Even as we have focused on different mathematical concepts with different groups of children, we have been consistently able to leverage this lesson in meaningful ways in our classrooms.

Here are some other places we have written about this work.

- Guzmán, L. D., & Craig, J. (2019). The world in your pocket: Digital media as invitations for transdisciplinary inquiry in mathematics classrooms. *Occasional Paper Series*, 2019(41), 6.

- Raygoza, M. C. (2016). Striving toward transformational resistance: Youth participatory action research in the mathematics classroom. *Journal of Urban Mathematics Education*, 9(2), 122-152.

- Thanheiser, E., & Koestler, C. (2021). If the world were a village: Learning mathematics while learning about the world. *Mathematics Teacher Educator*, 9(3), 202–228.

About the Authors

Courtney Koestler has been working in their current position at Ohio University since 2014, where they have had the opportunity to spend time in schools working alongside elementary school teacher-colleagues in their classrooms. They have been an educator since 1998, when they started their career as a middle school teacher, going on to work as a second-, fourth-, and finally fifth-grade classroom teacher before becoming a K-5 mathematics coach. It was in this role that they met administrators and colleagues who believed in child-centered and critical pedagogies, centered on taking an assets-based approach to teaching where students' (and families') interests, practices, and "funds of knowledge" were important resources on which to build and connect. Exposure to critical literacy and critical pedagogy made them question their teaching and realize they could be more intentional and purposeful about the kinds of lessons and pedagogical approaches they taught. Now, as a university-based teacher educator, they center equity and justice in teacher education and professional development work. Courtney recognizes that teaching is non-neutral; it is a political act.

Eva Thanheiser is a mother of three daughters and a professor of mathematics education at Portland State University. Eva has worked in elementary and middle schools regularly in a range of experiences from being a mathematics teacher for a fifth-grade class to running afterschool mathematics clubs. At the beginning of her career as a teacher educator, she focused mostly on content and motivation to learn. However, she learned at conferences about social justice, its inclusion into the mathematics classroom, and the power of that inclusion. She began trying out different activities and noticed that they allowed different kinds of participation. Over the years, she has tried more and more and kept reading up on tasks (books with tasks) and implementing them. She considers herself a lifelong learner and a contributor to improving struggles for social justice in mathematics education.

Mary Candace Raygoza is a STEMinist teacher educator at Saint Mary's College of California. Her scholarship explores teaching mathematics for social justice and critical, justice-oriented, anti-racist teacher education. Mary is a former high school mathematics teacher in East Los Angeles.

 **Jeff Craig** is committed to contemplating ethical questions in education. He also recognizes that children encounter ethical questions in their everyday life. In teaching, he reconciles ethical questioning against a backdrop of so-called Wicked Problems education, which prioritizes civic education as it relates to children as members of communities and societies. He finds the lesson presented here compelling because it engages children as both global and local thinkers who use mathematical and statistical techniques to understand their world and their positions within it.

 **Lynette Guzmán** is a mathematics education scholar who focuses on interrogating limiting discourses about people and their complexity. As a millennial who grew up with the internet, Lynette spends her time thinking about the ways digital platforms lend themselves to content creation, consumption, and remixing to promote particular kinds of discourses.

- I know my family and I do things the same as and different from other people and groups, and I know how to use what I learn from home, school, and other places that matter to me. (Identity 5)

- I like knowing people who are like me and different from me, and I treat each person with respect. (Diversity 6)

- I know when people are treated unfairly, and I can give examples of prejudice words, pictures, and rules. (Justice 12)

- I will work with my friends and family to make our school and community fair for everyone, and we will work hard and cooperate in order to achieve our goals. (Action 20)

MATHEMATICS CONCEPTS

- Understand a fraction as a whole partitioned into equal parts.

- Apply previous understandings of multiplication to multiply a fraction by a whole number.

- Solve word problems involving addition and subtraction of fractions.

- Draw graphs to represent data and solve problems involving addition and subtraction of fractions by using data presented in graphs.

LESSON 5.1 FAMILIES MATTER

Nicky Meindl

Social Justice Connection

In much of existing children's literature, family is portrayed as the nuclear family—made up of a cisgender husband, a cisgender wife, a cisgender daughter, and a cisgender son. This normative and dominant way of representing family structures is harmful and repressive to families that do not fit into this dominant perspective. This lesson explores the norms of family structures by asking students to examine family structures that they see both in their communities and the literature that they read. This lesson supports students' questioning about why these normative and dominant ways of representing family structures are often the only representations of families that students see in school curriculum. Students can take action toward diversifying the books in their school's library.

Deep and Rich Mathematics

Students collaborate to gather data on the representations of families and family structures found in their set of school library books and create a data display to explain their results to the class. At the conclusion of the lesson, the entire class will share their findings to create a larger data display of everyone's data. The teacher and students create a line plot that represents the diversity, or lack of diversity, of family representations found in the books. As students explain their results, the teacher prompts them to explain their noticings, wonderings, feelings, and desire for action.

RESOURCES AND MATERIALS

- Book: *My Friends and Me* by Stephanie Stansbie

- Video: Ms. Katie reading *My Friends and Me* (https://bit.ly/3y4o6zd) or an alternative video of Ali Ayars reading this book (https://bit.ly/3Dv3Wj7)

- Worksheet 1: *Family Type Tally Sheet*

- Teacher Resource 1: *Sample Letter*

MATHEMATICS PRACTICES

- Reason abstractly and quantitatively.

- Construct viable arguments and critique the reasoning of others.

Lesson 1 Facilitation

What Makes a Family?
LAUNCH (20 MINUTES)

- Begin with a think-pair-share. Ask students to answer these questions with a quick-write or drawing:

  + *What makes a family?*

  + *What does a family look like?*

  + *Who is in your family?*

- Next, have students share their responses in small groups. Walk around the classroom during these discussions to note what common and unique ideas arise among the students.

TEACHER NOTE

You should consider the social identities of the students in your classroom, such as if students are adopted or in foster care, when discussing families. Many times, these students have no idea when, or even if, they will see their birth families, so it is important to remind them of what makes a family and what ways their foster/adopted guardians care for them like "blood family." Find and read books throughout the year that show the foster care system to give students mirrors to see themselves in as well as windows for others to learn about a piece of their experiences.

- Create a class concept map connecting the ideas that students have about families (see Figure 1 as an example).

Figure 1. Families Concept Map

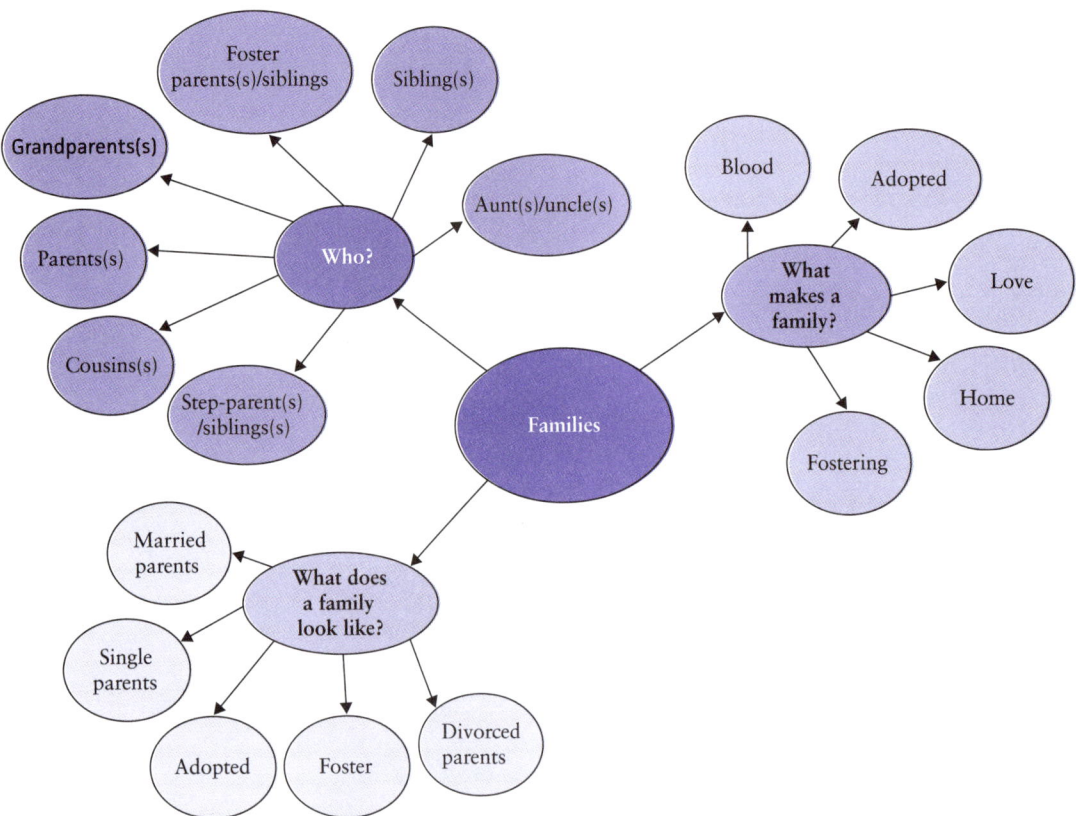

- Next, lead a discussion about the following questions:

  + *What families are usually represented in the books that you read?*

  + *What is your experience in reading books? Do you see your family represented in your reading?*

EXPLORE (30 MINUTES)

- Read *My Friends and Me* by Stephanie Stansbie (or watch a video of a teacher reading the book on YouTube; see the *Resources and Materials* section) to the class. As you read, stop and draw attention to the different appearances of families in the text. Here are some questions to ask:

  + *What do you notice about this family?*

  + *Is this family the same as the previous family?*

  + *What is similar? What is different?*

  + *What cultures do we see represented here?*

- When you reach the end of the book, aid students in creating a chart to describe the text's different family structures (see Figure 2 as an example).

Figure 2. Example Chart of Family Structures in *My Friends and Me* by Stephanie Stansbie

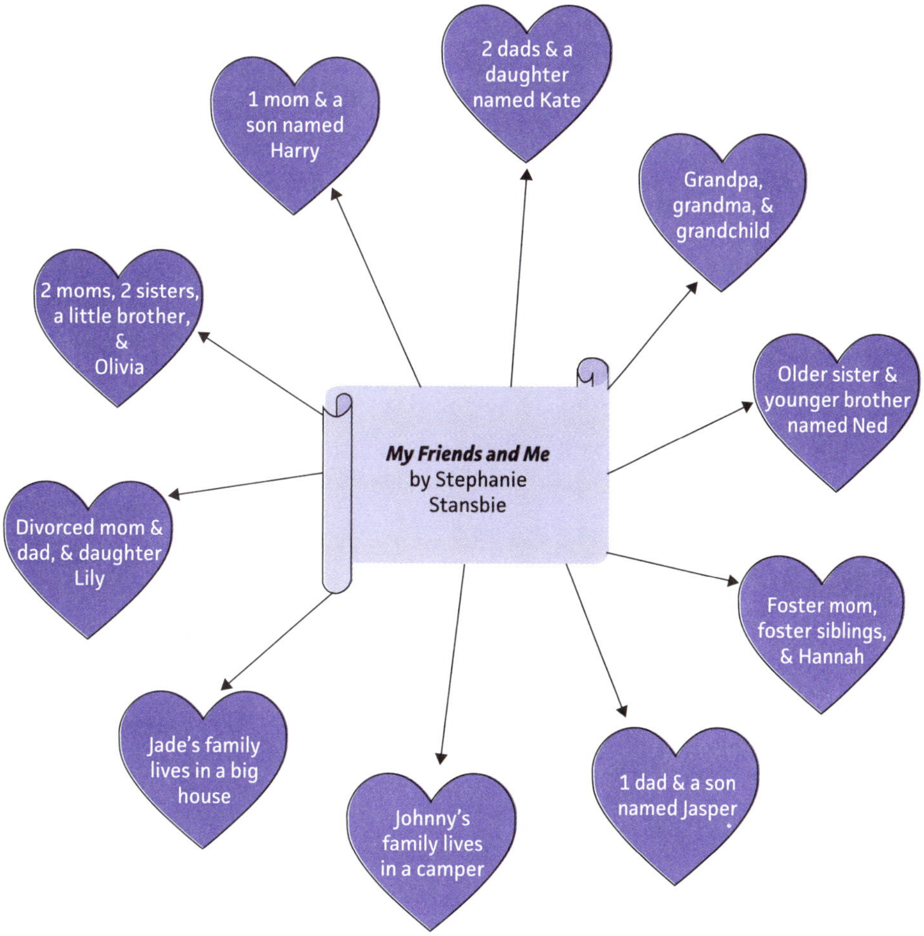

SUMMARIZE (15 MINUTES)

- Lead a whole-class discussion focused on these questions:

    + *What makes a family?*

    + *What are the different families we see represented in this story?*

    + *Who do we see in each of these families?*

    + *Do all families look the same? How do we know?*

    + *Does everyone in each family look identical?*

    + *Is every type of family represented in the book? How do we know?*

- Finally, draw students' attention to the wording on the second-to-last page, "Grown-ups . . . are fantastic at loving us." Ask students why they think the author chose to write "grown-ups" instead of "parents" here and how this is different from writing "parents are fantastic at loving us."

- Remind students to share a bit of new information that they learned in class with their families at home.

- As homework, have students find or create a picture of their family to share with classmates.

Lesson 2 Facilitation

Examining Our Library Books

Before this activity, you will confer with your school or local librarian to determine the extent to which the librarian will assist you in the lesson. The focus will be on identifying which books are checked out most often by students from the library. Students will work in small groups to analyze the family structures represented in each book, so picture books will allow students to work more quickly.

LAUNCH (10-15 MINUTES)

- Begin with this question: *How can we find out what types of families are represented in the books most read by students?* Allow students to share their ideas.

- Display the following goal on the board: "Gather and report on data of family structure representation in the most checked-out library books."

- Tell the class that today they will be looking closely at the pictures and mentions of families in this collection of books. Explain that when social scientists, such as sociologists, do this kind of work, they might go through the books a first time and develop their initial ideas about how to categorize the data; then they would meet and come up with a consistent approach they can use across all the different data sources, or books in this case. Tell the class that you will be doing the same thing. You will start by focusing on developing categories for types of families that everyone in the class can use. Then everyone will collect data for their books using those categories.

- Divide the students into small **heterogeneous groups** of three to four students.

- Each group will gather data from the library books about the family structures represented in some of the most popular (most checked-out) books in the library. If possible, the librarian can give a short presentation on the collection of most checked-out books.

- Tell the class that they will be looking closely at the pictures and mentions of families in the books for both primary and secondary characters. Tell the students that as they work, they need to come up with categories for the different types of families that they see in their books. Emphasize that they need the following for each category:

  + A definition or description (for example, "two moms" or "includes foster children")

  + An example from one of the books they are reviewing

Heterogeneous grouping of students—or grouping students with mixed abilities and strengths—is beneficial for students who historically have been underserved and marginalized in school as well as for those who consistently strive. It is beneficial for the learning of all students.

EXPLORE (45-60 MINUTES)

- Have students work in groups to develop categories. Circulate and pose questions such as these:

    + *What categories did you create?*

    + *Who did you include in your categories?*

    + *Who is missing from your categories?*

- When you feel groups have a rich enough set of categories, bring them together for a whole-class discussion.

SUMMARIZE (20-25 MINUTES)

- Tell the class that each group will share their ideas and then you will work collectively to develop a common list of categories everyone can use. But first emphasize that this can be a challenging topic by saying:

    Analyzing diversity can be tricky for us and for social scientists because we want to celebrate all kinds of diversity in families. We don't want to get too focused on separating people or families into groups. But without analyzing the types of families we see, it can be hard to notice patterns. So we will come up with a way to analyze the data, but we also need to remind ourselves that the important part is valuing all kinds of people and families.

- Have each group briefly share their categories (along with examples from their books) and record them on the board. Encourage groups to question and comment on each other's categories, considering why they created these categories, what might be missing from their ideas, noticing unique ideas, and talking about similar categories that could be combined.

- Lead the class in combining similar categories and developing a common list for each group to use as they work. One key issue to discuss with the class is whether a family is allowed to be listed in two places. For example, one family may include both married parents and foster children. You will likely need to allow for overlaps, as developing exclusive categories for types of families will be too limiting. This could also be a place to tie into the use of Venn diagrams.

- Conclude by having students share pictures of their families with one another.

- Remind students to share a bit of new information that they learned in class with their families at home.

Lesson 3 Facilitation

Representing Our Findings
LAUNCH (10 MINUTES)

- Remind the class of the categories developed in the previous lesson. You can distribute Worksheet 1 (*Family Type Tally Sheet*) so that students can list family types and record tallies. As a scaffold, you may want to prepare tally sheets that list the family types in a table so students can record their tallies as shown here:

Family Type	Character Type	Tally Marks	Frequency
Family type 1	Primary		
	Secondary		
Family type 2	Primary		
	Secondary		
. . .			

- Tell students that today they have two jobs, first to collect their data and then to create a graph of their group's data. Remind them that there are many types of graphs, and they need to choose one that they think most clearly communicates their data. **Reminder:** *Do not tell students what kind of graph to use; let them decide what would work best for them.* You may want to briefly discuss pie charts and how they require non-overlapping (exclusive) categories. Depending on past work in your class, you may want to review different types of graphs. You can choose whether you will have students make their graphs by hand or if they will have access to Google Sheets or other graphing software. If students are working by hand, graph paper is a helpful tool.

EXPLORE (30-45 MINUTES)

- Students work in their groups to create tally marks for each type of family that comes up in their books. If they encounter a family that they're not sure about, have them note where it is and save it so the rest of the class can discuss it after the data have been collected.

- When students are done, have groups discuss and then check in with you about how to represent their data with graphs. During this conversation, and later as you circulate and check in, pose questions such as these:

+ *Why did you choose this type of graph?*

+ *What are your labels?*

+ *What is your scale?*

+ *How does your graph tell a story about your data?*

SUMMARIZE (30-40 MINUTES)

- Have each group briefly present their graph to the class, answering the questions above.

- As a class, help the students combine their data sets and select one graph to make for the combined set of data. Discussion should focus on what graph they think will best represent the findings.

   Once the graph has been created, ask the students these questions:

   + *What does our class graph show us about the books in the library?*

   + *Does our class graph give us different information about the books in the library than each small group's graph? Why or why not?*

   + *Based on the information in our class graph, do we think that the most-often checked-out books in the library have a diverse representation of family structures? How do we know?*

   + *What types of families are well represented? What types are not well represented?*

   + *Why is there (or is there not) a diverse representation of family structures in the most checked-out books?*

   + *What can we do to have more diverse family structures represented in our library books? What can we do to encourage students to check out these books?*

- Remind students to share a bit of new information that they learned in class with their families at home.

Lesson 4 Facilitation

Connecting Research to Action

This lesson provides an opportunity for students to connect their research to action. You can decide if you want to have different students or groups to take different forms of action or if you want to plan a common action (such as a presentation and/or letter-writing campaign) that everyone will participate in.

LAUNCH (10-15 MINUTES)

- Ask the class to brainstorm ideas for how they can take action based on the research they have done so far. Here are some possible ideas:

    + Running a donation drive for books with diverse representations of families

    + Sharing results on social media and soliciting suggestions for book recommendations

    + Sharing results with family members

    + Sharing results with and/or writing letters advocating for change to the librarian, principal, school board, parent/caregiver teacher association (PCTA), or other members of the community

EXPLORE (35-45 MINUTES)

- Students work on their selected form of action. Here we provide directions for a letter-writing campaign. The students will draft letters to their school's librarian, principal, PCTA, school board members, and/or district members. See Teacher Resource 1 for an example. The letters will advocate for more inclusive literature to be added to the school library by responding to the following questions:

    + *What is the importance of having diverse family structures represented in books?*

    + *What is the importance of adding more inclusive books to the library?*

    + *What are some books that could be added to the library? How do these books diversify the library through family structures?*

- The letters should be drafted, peer-edited, typed, and sent to their respective recipients.

SUMMARIZE (15-20 MINUTES)

- Groups of students share about how they connected research to action, why they chose that approach, and what they learned from the experience.

- Remind students to share a bit of new information that they learned in class with their families at home.

TAKING ACTION

As part of the end of the lesson, the students will write letters to the school librarian, school principal, school board members, or school district representatives to advocate for the addition of books that reflect the diverse family structures in both their community and their world. The students will detail why adding these books are important, provide suggestions for one or two books to be added to the school's library, and explain their reasoning for selecting the book.

As an extension activity, the students could work with their local and school libraries to highlight books that show diverse family structures. The students would be working with their librarians to create short summaries or book reviews to help entice other readers to select, read, and then share the book with others in their community. This could also include small art projects to create visuals to display about the highlighted books with diverse family structures, which would help draw attention to these books.

Communicating With Stakeholders

At the end of each day, the students will be prompted to share a bit of new information that they learned in class with their families at home. Additionally, the students' homework will be to bring in a picture or drawing of their family from home to share with the class in small groups, which honors the diverse family structures that are represented in the classroom community as well as helps students make connections between the similarities and differences among their families. Encourage families and parents/caregivers to visit the classroom to share their family's history and experiences. This connects to the Identity and Diversity domains of the Social Justice Standards, as students will be learning about their peers' families and comparing this to their own families.

Online Resources

 Available for download at **https://qrs.ly/wbfixtr**

Family Type Tally Sheet

Family Type	Character Type	Tally Marks	Frequency

▲ *Worksheet 1: Family Type Tally Sheet*

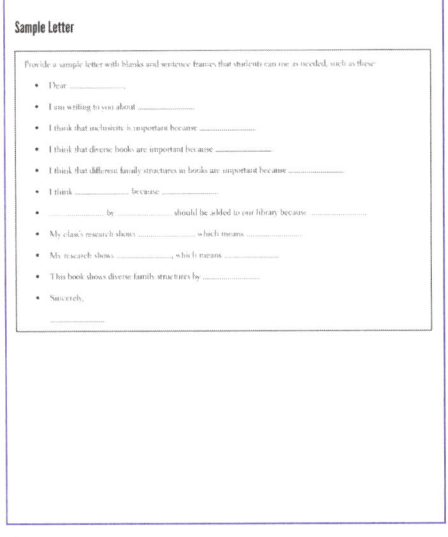

▲ *Teacher Resource 1: Sample Letter*

About the Author

Nicky Meindl is a queer, Chicanx, and nonbinary, first-generation graduate student in the Curriculum and Instruction Master's program at Chapman University while simultaneously student teaching with the purpose of becoming a social justice elementary educator. Their research focuses on the intersection of Queer Theory and Ethnic Studies in elementary schools.

LESSON 5.6 CHALLENGING ABLEIST ASSUMPTIONS IN MATHEMATICS PROBLEMS

Courtney Koestler, Jennifer R. Newton, and Jan McGarry

Social Justice Connection

This lesson explores the issue of human diversity (i.e., different kinds of bodies), disability, and ableism. This lesson is meant to be launched during or after a typical lesson found in many textbooks that assumes students all have "typical" bodies, such as having 10 fingers, and are able to participate in "typical" ways. Students can use critical literacy skills to examine the mathematics lesson as presented as usual (in many textbooks) as well as resources in their classroom to see how bodies, disability, and ableism are presented. Oftentimes the topic of disability in mainstream classrooms is invisible or explicitly not talked about unless absolutely necessary, and it is important for students to see both children and adults with disabilities represented in their classrooms through empowering ways. Disabilities should be portrayed in ways that avoid deficits and stereotypes and instead accurately describe the disability and/or portray people with disabilities living their lives (whether or not disability is the focus).

SOCIAL JUSTICE OUTCOMES

- I like knowing people who are like me and different from me, and I treat each person with respect. (Diversity 6)

- I have accurate, respectful words to describe how I am similar to and different from people who share my identities and those who have other identities. (Diversity 7)

- I feel connected to other people and know how to talk, work, and play with others even when we are different or when we disagree. (Diversity 9)

MATHEMATICS CONCEPTS

- Adding and subtracting multiples of 10 based on place value and properties of operations.

- Multiplying whole numbers by multiples of 10.

- Understanding and making generalizations about place value; specifically, understanding and justifying that a digit in one place represents 10 times what it represents in the place to its right.

MATHEMATICS PRACTICES

- Construct viable arguments and critique the reasoning of others.

- Look for and make use of structure.

- Look for and express regularity in repeated reasoning.

Deep and Rich Mathematics

This lesson engages students in using their bodies (i.e., their fingers) as a physical representation to support skip-counting groups of 10. At the same time, students will also unpack this common practice to begin a conversation about body diversity.

RESOURCES AND MATERIALS

- Chalkboard/whiteboard, chalk/markers, or someplace else to record students' thinking

- Choose among the following books to include in your classroom library:

Picture Books

- *All Are Welcome* by Alexandra Penfold

- *The Bug Girl* by Sophia Spencer

- *Emmanuel's Dream: The True Story of Emmanuel Ofosu Yeboah* by Laurie Ann Thompson and Sean Qualls

- *Hello Goodbye Dog* by Maria Gianferrari

- *I Am Not a Label* by Cerrie Burnell

- *A Kids Book About Disabilities* by Kristine Napper

- *Mama Zooms* by Jane Cowen-Fletcher

- *Rescue and Jessica: A Life-Changing Friendship* by Jessica Kensky and Patrick Downes

- *Terry Fox and Me* by Mary Beth Leatherdale

- *What Happened to You?* by James Catchpole and Karen George

Chapter Books

These may be more appropriate for older grades.

- *Braced* by Alyson Gerber

- *Roll With It* by Jamie Sumner

- *Intersectional Allies: We Make Room for All* by Chelsea Johnson, LaToya Council, and Carolyn Choi

Reference Books for You

- *Critical Literacy Across the K-6 Curriculum* by Vivian Maria Vasquez

Additional Resources

- Article: "How to talk to your kid about disabilities," by Caroline Bologna, *Huffington Post*, March 1, 2021 (https://bit.ly/32Svi68)

- Lesson: Learning for Justice, "Picturing Accessibility: Art, Activism and Physical Disabilities" (https://bit.ly/3oeLVkC)

- Lesson: Learning for Justice, "What Is Ableism?" (https://bit.ly/3oe0kh9)

- Lesson: Learning for Justice, "What Is a Disability?" (https://bit.ly/3lrFFUG)

Lesson 1 Facilitation

Introducing Assumptions

LAUNCH (30 MINUTES)

- A common elementary mathematics activity is to count by tens by counting all the fingers in the classroom. This implicitly makes an ableist assumption that all people have 10 fingers. This lesson engages students in questioning those assumptions and thinking about different mathematical contexts we can use for making tens. Prior to implementing this lesson, we recommend you consider your classroom and how you want to frame the topic. This is important for all classes, but it is especially critical if you have any students who do not have 10 fingers. The *Resources and Materials* section has several links that can be helpful in thinking about how to discuss these topics with your students. If you have a student(s) who does not have 10 fingers, we suggest you discuss the lesson with the family and child to make sure you are approaching it in a way that feels inclusive and supportive to the student(s). While we provide suggestions for some possible approaches to navigating this space, you should adjust based on your context. Say something to students like this:

  I have been thinking a lot about today's activity and wanted to talk to you about it. It is an activity that is in a lot of textbooks because it is usually really good at getting kids to think about patterns in our number system, but I also am wondering about some assumptions it makes about kids and bodies. Let's look at the task and think about some of the hidden assumptions it makes before we start.

TEACHER NOTE

You should adjust what you would say depending on your context. For example, if you have already done work with critical literacy, your students may be familiar with the idea of how authors' assumptions and biases can be analyzed. If not, you may have to discuss it a bit more, perhaps by asking if they know what an assumption is and if they can give any examples. For more information, see Vivian Maria Vasquez's (2016) book, *Critical Literacy Across the K-6 Curriculum*.

- Say to students: *The activity that is usually in textbooks is to "figure out the total number of fingers in our class by counting by tens."* (You may want this displayed on the board as well.) If students need additional support, you may want to explicitly ask:

  + *What does this problem assume about people's bodies? About their fingers?*

  + *How do you know the problem assumes this?*

- Have students share their responses and ask the class: *Does everyone in our world have 10 fingers? Why or why not?* Students may have examples of people in their lives who were born with a different number of fingers or who lost a finger(s) during their lifetime. If not, explain that this could happen. If you have a student who does not have 10 fingers in your class, be mindful of not expecting them to speak as a "representative" for those with a different number of fingers and of not allowing other students to do the same.

- At this point, we offer two possible ways to move forward in the lesson. You may choose to continue with the activity as it is commonly stated (count by tens to find the total number of fingers in the class), or you may ask the class to brainstorm other things that come in tens that you can use for the problem (we especially recommend this second approach if someone(s) in the class does not have 10 fingers). You might tell the class: *The mathematics goal for this problem is to count by tens, but since not everyone has 10 fingers, let's think of some other things that could come in tens that we can count by instead.* The class might suggest ideas like boxes of markers, bags of marbles, or the tens sticks from base-10 blocks. Choose one of these, and prepare one for each person in the class plus several extras before continuing the lesson. Alternatively, you can have the base-10 sticks available and suggest to the class that you use these to represent whatever group of 10 objects the class has agreed on. Give each student one of the tens.

- Tell the class, *We're going to count by tens to see how many ___ we have in our whole class.*

- Have students raise their hands or objects as they say their numbers (10, 20, 30 . . .). You can keep track of the count by listing the numbers, drawing a number line, or using other ways your students suggest to record the numbers.

EXPLORE (30-45 MINUTES)

- Try it more than once, starting with different students. Ask students to describe what they notice. Ask students questions about what would happen in different scenarios. Here are some examples:

  + *If there are any students absent from class, what would be the number of [objects] if they were present?*

  + *What if the art, music, and physical education teachers joined the class?*

  + *What if we had 35 students in our class?*

  + *What if we counted the whole third grade, which has 78 students?*

- Be sure to entertain multiple strategies for solving the tasks (e.g., counting by tens, adding on by the multiple of 10). As you work, record the problems and solutions (e.g., $18 \times 10 = 180$) in a vertical list off to the side or on chart paper that you can refer back to in the next lesson when the class will look for patterns.

SUMMARIZE (20 MINUTES)

- Briefly summarize the mathematical strategies that students used or ask them to identify any common strategies. Then say that you'll be exploring problems with tens more next time. Next, read one of the suggested children's books. It may be good to have several books to choose from and to begin a study of how disability is explored similarly and differently in the books. We suggest that you continue to have books like those suggested in the *Resources and Materials* in your classroom throughout the school year so that students see these are a part of the classroom library, and not just as part of a lesson on disability.

- While you are reading and/or once you are done reading, have students share questions they have about differences and disabilities.

Lesson 2 Facilitation

Exploring Place Value

LAUNCH (20 MINUTES)

- Remind the class of your work last time, both about questioning the hidden assumptions in mathematics problems and your mathematical strategies. Explain that today you will focus more on the mathematical strategies and in a future lesson you will return to exploring hidden assumptions and inclusion in mathematics problems.

- The list of problems from Lesson 1 (e.g., $18 \times 10 = 180$) should be displayed to the side of the board, but do not focus on it yet. Pose some questions like the following to the class: *How many tens does it take to make 780? How do you know?*

- Review the meaning of the place value for a number like 780. What does each digit stand for? Point out that the 7 stands for 7 *hundreds*, and ask them how many *tens* we can think of this as. Point out the connection to the previous problem: 780 stands for 7 hundreds and 8 tens, but it can also be thought of as 78 tens.

- Review the list of problems solved from Lesson 1 (e.g., $18 \times 10 = 180$) and the list of problems solved today. Ask the students if they notice any patterns. You can also ask more explicitly: *What happens when we multiply a number by 10?* Students will likely notice that you "add" a zero to the end of the number. You should question or clarify the use of the word "add" and point out that they don't mean, for example, $18 + 0$. You can suggest that another word they can use is "append" or that we might think of this as moving the digits one place to the left (i.e., ones move to the tens place, tens move to the hundreds place, and so on).

- Write this as a class conjecture. If students are not familiar with that term, explain that it's something that you think is true in mathematics, but you need to explore it more to determine if it's always true or not. Tell the class that today they will be working with examples, base-ten blocks, and pictures to try to determine if this is always true and to then prove that it's true.

EXPLORE (20 MINUTES)

- Students work in pairs or small groups to explore the conjecture. They may want to begin with several examples, then ask them to think about how they can show that it would work with any numbers. Encourage them to think about how they can use base-10 blocks and/or drawings to show what is happening and why that makes this true. Question groups about how they can use what they learned from the warm-up to help them on this problem (e.g., that 780 has 78 tens).

- It may help to have students first think about single-digit numbers (e.g., 7 × 10) before thinking about double-digit numbers. Circulate among the groups, see what they are thinking, ask probing questions, and make decisions about what explanations to have students share and in what order.

OPTIONAL EXTENSIONS

It is likely that most students will implicitly stick to whole numbers. Depending on the progress and understanding of the class, you may choose to challenge individual groups or the whole class with different types of numbers such as fractions or decimals. This can be helpful for a conversation about how to modify and/or limit conjectures. The conjecture could be limited to whole numbers, or it could be modified to include decimals by being precise about shifting digits one place value to the left as opposed to appending a zero. The conjecture does not apply to fractions. You can also explore what happens when we *divide* a number by 10.

SUMMARIZE (30 MINUTES)

- Have some or all groups share their thinking about why multiplying by 10 results in appending a zero to the end of the number. Before starting, tell students that when they listen to groups share, they should think of compliments (i.e., something they found valuable about the group's *mathematics*) and questions (i.e., things they didn't understand or want to understand better). The class should engage in a discussion around the different explanations.

- End by highlighting the key strategies that students used and emphasize that mathematicians often spend a long time developing a clear proof that something is true, so it's not something you usually figure out in one day. Depending on the progress made, you may choose to revisit this concept in the future.

- Explain that next time you will return to a focus on the hidden assumptions in mathematics problems.

TAKING ACTION

Remind the class that this lesson began by identifying the hidden assumptions in the common mathematics task: *Figure out the total number of fingers in our class by counting by tens*. Read the book, *Intersectional Allies: We Make Room for All*, by Chelsea Johnson, LaToya Council, and Carolyn Choi. As with any book, you should prepare by prereading *Intersectional Allies*, as there is text in other languages.

Discuss generally what it means to be an ally to others, asking students what they think the word means and asking for examples of allyship. For example, you may ask: *What are the ways that the children in the book acted as allies for their friends? What are ways families supported other families?*

Next, discuss how allyship was framed in the books you used in the previous section. Were there friends, adults, and others that offered supports and accommodations that provided access to people in the books? In what ways?

Taking Action Option 1

Ask students if they know of examples of supports and accommodations that provide access to people with disabilities at their school or in public buildings (e.g., automatic door openers, braille lettering on signs, accessible parking, accessible restrooms).

Ask students to analyze the ways in which the school building is welcoming and safe for different kinds of people, especially for those with different kinds of disabilities. If possible, invite a guest speaker, such as a local disability advocate, to collaborate.

If or when students find issues with accessibility, support them in taking action by communicating via letters or a presentation with building and district administration, school board members, and community members.

Taking Action Option 2

As an ongoing investigation, have students examine ways in which people are portrayed in the books in your classroom, including in your mathematics curricular materials.

For example, you may compare how people with disabilities are represented in *All Are Welcome* by Alexandra Penfold (i.e., where the children just happen to be using a wheelchair or a white cane but not specifically discussed as having disabilities) versus in *Emmanuel's Dream: The True Story of Emmanuel Ofosu*

Yeboah by Laurie Ann Thompson and Sean Qualls (i.e., where his true-life story is illuminated about what it was like growing up with a disability).

Students can note places in their textbooks where there are assumptions that everyone in the classrooms is the same, especially in terms of being able-bodied. They may choose to take action by writing letters to different audiences, such as the textbook publishers or district administrators (curriculum coordinators), to describe their findings, let them know how this is not an accurate depiction of people in their classroom and/or world, and give suggestions of ways to make the task or lesson more inclusive. While this activity is ideally student-led, students may need some assistance in developing more inclusive tasks.

Communicating With Stakeholders

Before teaching this lesson, you should reach out to families, parents/caregivers, and also administrators in your building to provide an overview of the topics included in this lesson (different kinds of bodies, disability, and ableism) and the kinds of the discussions that might emerge. This will help you anticipate ways to be sensitive to and inclusive of the students in your classroom. Any information you receive about specific students or their family members (about differences or disabilities) should stay private, unless they give you explicit permission to share. And, as mentioned earlier in the lesson, take care not to place any student(s) in a position where they have to speak as a "representative" for those who are different or who are disabled.

About the Authors

Courtney Koestler is a proud former public school teacher and currently serves as the Director of the OHIO Center for Equity in Mathematics and Science in the Patton College of Education at Ohio University. Their work centers on critical literacy and critical pedagogies in early childhood and elementary education.

Jennifer (Jen) R. Newton began her career as an inclusive early childhood educator in 2000, and has worked across states and settings to promote inclusive practices for students with disabilities. The opportunity to collaborate with Courtney has enabled her to advance anti-racist and anti-ableist work with teacher candidates and teacher education broadly.

Jan McGarry is an elementary teacher in Athens, Ohio. She has had the privilege of working with first and second graders in Appalachia for 20 years. Jan has a passion for fostering inclusive classroom families that center students' voices and encourage connections with the community and current events through the lens of social justice education.

LESSON 7.2 THE TRUE COST OF THAT $29 T-SHIRT IN THE STORE WINDOW

Bethany Chan, Debasmita Basu, Rebecca Ellis, Frances K. Harper, and Jennifer Ruef

Human Rights

To lower their production cost and maximize their profit, companies often establish their manufacturing divisions in cheaper and less regulated locations such as Bangladesh, China, Mexico, Cambodia, and other developing countries. The working conditions of the factories, commonly known as sweatshops, are often hazardous and abusive with long working hours and inadequate pay. For example, despite fashion being a $29 billion USD industry, people working in garment factories in Bangladesh are only paid $0.35 USD an hour, which forces them to work for 14–16 hours a day to pay for their daily necessities. In 2013, one such factory on the outskirts of Dhaka, Bangladesh, collapsed, trapping and killing more than a thousand of its employees. The investigation suggested that the factory was under scrutiny because of evidence of unsafe conditions, but no steps were taken to improve them. The goal for this lesson is for students to use mathematics in this social justice context to raise their consciousness about the exploitation being practiced within the four walls of the sweatshops. This lesson gives students a chance to critique the world and explore their own identities, without teachers or other adults imposing their own beliefs.

- Make sense of problems and persevere in solving them.
- Reason abstractly and quantitatively.
- Construct viable arguments and critique the reasoning of others.
- Model with mathematics.
- Look for and make use of structure.
- Look for and express regularity in repeated reasoning.

Deep and Rich Mathematics

This lesson is designed for students to apply prior knowledge of percentages and proportions to construct and operate on mathematical models based on real-life scenarios from the clothing industry. Students will use critical thinking skills and problem-solving strategies as they apply and attain mathematical content knowledge. For this reason, teachers should be prepared to share just-in-time information about problem-solving methods others may have used. This preparation will support teachers in recognizing solution strategies and potential challenges in mathematical understandings and preparing to ask focusing questions to guide students' work. Teachers are invited to scaffold the lesson based on anticipation of their students' prior knowledge and experience.

About the Lesson

This lesson uses a launch–explore–summarize instructional model and is intended to take approximately 120 minutes to complete across two class periods.

Lesson 1: Students use ratios, proportions, and percentages to estimate the allocation of money to make a t-shirt.

Lesson 2: Students use ratios, proportions, and percentages to analyze the actual profit allocation of a t-shirt and consider what a fair allocation would be.

RESOURCES AND MATERIALS

- Worksheet 1: *How Should Money Be Allocated?* (1 per student)
- Worksheet 2: *How Is Money Allocated?* (1 per student)
- Worksheet 3: *What Is Fair?* (1 per student)
- Highlighters, colored pencils, or crayons
- Video: "True Cost Clothing Industry" (https://france-sharper.com/clothing-industry-video/)
- Website: Educational Video Center (https://evc.org/)
- Website: Two Dollar Challenge (http://twodollarchallenge.org/our-story/)

Lesson 1 Facilitation

How Do You Think Money Is Allocated?
LAUNCH (25 MINUTES)

- Assign stores to students to research the cost of clothing items. Students can bring in advertisements from clothing retailers or look online. Have students share what they found.

  Note: *If students find an item that costs around $29, you might highlight that item for the lesson later on.*

- Have students brainstorm ideas about *what the money pays for* when they purchase an item of clothing. Students might find it helpful to pick one of the items of clothing that they found in advertisements or online. The point of this activity is to get students thinking about the production process, costs, and so on.

- Elicit some ideas from students and then show the video, "True Cost Clothing Industry" (https://francesharper.com/clothing-industry-video/).

- Ask students to reflect on what they noticed or wondered as they watched the video. Then have them share out with the class and make sure students understand the eight different categories involved in the production process. Consider asking the students the categories they remember and writing the list publicly on the board. Thus, students can refer to the list when answering question 1 on Worksheet 1 (*How Should Money Be Allocated?*).

EXPLORE (35 MINUTES)

- Provide students with Worksheet 1. Ask:

  + *Who and what is involved in the t-shirt-making process?*

  + *How do you think the money is allocated among the eight categories?*

- Tell students to individually complete question 2, which requires them to color/shade the part of the dollar bill that correlates to each category.

- After they complete question 2 and distribute the $100 bill among eight categories, put students in groups of four.

- Ask the students to share their strategies with their group members and discuss the rationale behind the money allocations they choose.

- From each group, ask one student to volunteer for the role of group leader to share what they discussed in their group during the whole-class discussion.

SUMMARIZE (10 MINUTES)

- Facilitate a whole-class discussion. Ask students:

  + *If the $100 bill represents the retail price of a product, how do you think that the $100 bill is distributed among the eight categories? Why?*

- For students who divide the bill nonuniformly, ask:

  + *Who should get the maximum portion of the money? Why?*

  + *Who should get the minimum portion of the money? Why?*

Lesson 2 Facilitation

How Is Money Allocated? What Is Fair?
LAUNCH (10 MINUTES)

- Distribute Worksheet 2 (*How Is Money Allocated?*). Lead a brief discussion about how money for an item of apparel is typically distributed. **Note:** *Values are included on the worksheet.*

  ☐ *Materials: 12%*   ☐ *Retailer: 58%*

  ☐ *Factory profit: 4%*   ☐ *Intermediary costs: 4%*

  ☐ *Transport: 8%*   ☐ *Brand profit: 12.4%*

  ☐ *Overhead cost: 1%*   ☐ *Workers: 0.6%*

- Individually, or in pairs, have students color in the $100 bill using the new breakdown.

EXPLORE (35 MINUTES)

- Divide students into groups of four and assign each group to one of the two group types (see the following table). Students in each group type will focus on some of the percentages given in the Lesson 2 Launch and will calculate how the cost of a $29 t-shirt is distributed across the different categories. Students should each take on one conversion individually and then share their strategies with the other group members.

Group Type 1	Group Type 2
Retail	Intermediary
Profit to brand	Factory profit
Material cost	Overhead cost
Transportation	Payment to the workers

- After students share their strategies within their group, pair up group types and encourage them to compare the amount of money distributed among the eight categories.

- Begin a whole-class discussion by asking, *How does this breakdown differ from what was imagined on Worksheet 1?*

- Use the following questions to further assess students' understanding:

  + *Which among the eight categories received the highest percentage of the retail price of the t-shirt? How much is allotted to that category?*

  + *Which among the eight categories received the lowest percentage of the retail price of the t-shirt? How much is allotted to that category?*

  + *How many times more did the retailer earn compared to the workers who are employed to produce the t-shirt?*

  + *How do the numbers look when we are talking about portions of $29 instead of percentages?*

- Tell students to return to the groups they were in for Worksheet 1 and distribute Worksheet 3 (*What Is Fair?*).

- Have students work in groups as they complete the worksheet and record their reasoning for each question.

- As students work, encourage them to think about both the context and the mathematics as they create their responses.

- Use purposeful selection and sequencing to have students share the key ideas from the questions on Worksheet 3 during a whole-class discussion. Pay particular attention to the students' answers to question 4, as this will guide the transition to the summary portion of the lesson.

SUMMARIZE (15 MINUTES)

- Use the following questions to facilitate a closing discussion:

  + *Other than workers' wages, what other factors impacting the workers should companies consider? What benefits can companies get from being ethical?*

    - Possible Answers: *A positive company image attracts more customers/customer loyalty. It creates a company culture that values being ethical and socially responsible. These companies provide jobs to the local community by not outsourcing.*

- Continue the discussion by saying, *Some students have suggested buying only from ethical companies, but that can get expensive. For families that cannot afford to pay more money for ethically made clothing, what are other actions they can take to fight against unethical companies?*

- Have students brainstorm ways they can fight (not support) sweatshops and the unfair treatment (monetarily, physically, and mentally) of workers in the clothing industry.

TAKING ACTION

- Have students research current companies that sell sweatshop-free clothing (e.g., Patagonia) and create a brief presentation to share with your community (e.g., other students at the school, community members, family).

  - Consider the following questions: *Beyond working conditions, what other questions do you have about these companies? Do sweatshop-free clothing companies distribute money differently and/or fairly?*

- Brainstorm individual and family actions to take in relation to ecojustice (e.g., buy used clothing, buy local, wear clothing until it wears out, repair clothing when possible, make your own clothes).

- Create a video or podcast in which you interview sweatshop workers, organizations, or social justice members who have worked toward gaining more rights for workers. Share your video or podcast with the community. See the Educational Video Center (https://evc.org/).

  - Another possible topic: history of labor laws

- Create a list of questions to ask and find out if local businesses or community members endorse sweatshop-free products. If they don't, brainstorm ways to appeal to the community on the importance of supporting sweatshop-free products.

- Challenge families, community members, other teachers, and others to the Two Dollar Challenge (live on $2 for 5 days; http://twodollarchallenge .org/our-story/) as a way to raise awareness of global poverty and raise money for an organization working for social justice.

Online Resources

[online resources] Available for download at **https://qrs.ly/wbfixtr**

▼ *Worksheet 1: How Should Money Be Allocated?*

Name: _____

How Should Money Be Allocated?

Imagine that you buy an item of clothing from the store for $100. How should the money be allocated?

1. Who and what is involved in the t-shirt-making process? Create a key matching a color to each category involved in the production process that we discussed as a class.

   Key:

2. The $100 bill below has been subdivided into 100 equal sections, where each section is worth $1. How do you think the money is divided among the eight categories?

   a. Divide the bill among the categories by coloring in the sections on the $100 bill.

3. Discuss how you divided your bill with your group.

   a. Are you in agreement with what others said?

   b. What differed?

   c. If you could change anything in your drawing, what would you change?

   Summarize your discussion here.

▼ *Worksheet 2: How Is Money Allocated?*

Name: _____

How Is Money Allocated?

1. Your teacher has now shown you how money from an item is typically divided. On the $100 bill below, show how the money is actually allocated.

   ☐ Materials: 12%   ☐ Retailer: 58%
   ☐ Factory profit: 4%   ☐ Intermediary costs: 4%
   ☐ Transport: 8%   ☐ Brand profit: 12.4%
   ☐ Overhead cost: 1%   ☐ Workers: 0.6%

2. If a shirt costs $29, and not $100, how much money is allocated to each category involved in the production process based on the percentages in Question 1?

3. Compare and contrast how the money is allocated in your models (Worksheet 1) versus in the model above (Question 1).

▼ *Worksheet 3: What Is Fair?*

Name: _____

What Is Fair?

In Worksheet 1, you modeled how you think money is divided.

In Worksheet 2, you modeled how money is typically divided.

Now, think about what is *fair*. How should money be allocated?

1. Divide $100 among the eight categories in a way that you think is fair. Color in the bill and label the key.

   ☐ Materials: _____%   ☐ Retailer: _____%
   ☐ Factory profit: _____%   ☐ Intermediary costs: _____%
   ☐ Transport: _____%   ☐ Brand profit: _____%
   ☐ Overhead cost: _____%   ☐ Workers: _____%

2. Why do you allocate the money in this way?

3. How does this new allocation differ from the previous two versions?

4. What questions do you have about allocating money among the eight categories involved in the production process?

- I can recognize and describe unfairness and injustice in many forms including attitudes, speech, behaviors, practices, and laws. (Justice 12)

- I know about some of the people, groups, and events in social justice history and about the beliefs and ideas that influenced them. (Justice 15)

- I will speak up or take action when I see unfairness, even if those around me do not, and I will not let others convince me to go along with injustice. (Action 19)

ESSENTIAL MIDDLE GRADES CONCEPTS

- **Statistics and Probability—** Grade 8: Investigate patterns of association in bivariate data.

MATHEMATICS PRACTICES

- Reason abstractly and quantitatively.

- Construct viable arguments and critique the reasoning of others.

- Model with mathematics.

- Use appropriate tools strategically.

LESSON 9.4 HOW MANY MEALS CAN MINIMUM WAGE BUY?

Elizabeth O. Ayisi and Colleen Carman

Economic Inequality

Young adults are often blamed for their debt, but is it really their fault? What about people living in poverty? This lesson examines the relationship between cost of living and minimum wage over time. Students will be able to determine whether their dollar amount really goes as far as it should. It won't be long until these young adults are applying for jobs, deciding whether to go to college and where, and/or determining how to budget their finances. This lesson may spark the debate over whether these choices will be enough for them to be financially comfortable, to reduce poverty in America, and to extend or develop how to empower themselves and other young adults.

Deep and Rich Mathematics

Students will learn to build and analyze scatter plots while also estimating the line of best fit. They will conceptualize and analyze the slope, positive and negative association, and linear and nonlinear correlation. Additionally, students will calculate several slopes, compare slopes, and discuss the meaning of the slopes in a real-world context.

About the Lesson

This lesson uses a launch-explore-summarize instructional model and is intended to take approximately 165 minutes to complete across three class periods.

Lesson 1: Students investigate the cost of living to answer the question: *Can you buy more Big Macs with an average salary in 2020 or with an average salary in 1968? Why?*

Lesson 2: Students compare the average rate of change of minimum wage and cost of living to understand what the slope of each line (minimum wage and cost of living) represents and how the steepness of each line affects affordability.

Lesson 3: Students will create a presentation of their argument about when they think Big Macs were more affordable.

RESOURCES AND MATERIALS

- Teacher Resource 1: *Exit Slips* (1 per student)

- Worksheet 1: *Minimum Wage and Cost of Living* (1 per student)

- Worksheet 2: *Federal Minimum Wage* (1 per student)

- Worksheet 3: *Reflection* (multiple copies per student)

- Student access to computer with internet

- Website: Desmos.com graphing calculator

- Website: American Institute of Economic Research online cost of living calculator (https://bit.ly/31sLlXG)

- Website: FRED economic data, "Federal Minimum Wage Graph" (https://bit.ly/3I9ZDxf)

- YouTube video: "When Can You Afford More Big Macs? In 1968 or Now?" TikTok video by @genypastor (https://bit.ly/3rx4geE)

- Student access to Google Slides, PowerPoint, chart paper, or another tool to present their arguments

Lesson 1 Facilitation

Minimum Wage and Cost of Living

Students will watch a TikTok video that introduces the question: *Can you buy more Big Macs with an average salary in 2020 or with an average salary in 1968? Why?* Students will then investigate the cost of living over this time frame through analysis of slope. Students will then make connections and prepare a presentation through research of historical events that may have impacted the cost of living.

LAUNCH (10 MINUTES)

- Distribute Worksheet 1 (*Minimum Wage and Cost of Living*).

- Spark curiosity by showing the video, "When Can You Afford More Big Macs? In 1968 or Now?" (https://bit.ly/3rx4geE) and have students jot down some initial thoughts and observations on the worksheet (question 1.a).

- Present the Federal Minimum Wage graph on Worksheet 1 and have students label and analyze the graph.

  + Ask about units for *x*- and *y*-values and what these values represent.

  + Make a list of observations students can see. Focus on using precise vocabulary.

  + Notice that the data points on the graph are from the years 1968 to 2020.

- Tell students to make a hypothesis based on the video and the graph.

  + Encourage them to list what they notice and wonder.

  + Take a poll for each option and ask a few students to explain which option they chose and why.

  + Encourage students to think about whether wage is the only important factor in the affordability of Big Macs. Make a list of the factors students share.

EXPLORE (35 MINUTES)

- Present the Cost of Living graph on Worksheet 1 and have students label and analyze the graph.

  + Ask students if they know what the cost of living means.

  + Ask about the units for *x*- and *y*-values and what these values represent. Have students describe what one of the (*x*,*y*) points represents.

- - Make a list of students' observations. Focus on using precise vocabulary.

  - Notice that the data points on the graph are from the years 1968 to 2020.

- Introduce the Online Cost of Living Calculator and discuss what information it gives and what each value represents.

- Tell students to work in their groups to complete Worksheet 1.

- While monitoring group work, ask students:

  - *What are the units of your rates of change?*

  - *What do your results mean?*

- Regroup when most groups have finished question 5 to highlight student work, including different representations and methods to find the slope between data points. Use purposeful selection, sequencing, and connecting techniques to call on groups to present their ideas as a class.

- Use the following questions to assess and facilitate student understanding of slope and context:

  - *When is it best to use each method?*

  - *Which slope was higher?*

  - *What does that mean in terms of buying power?*

  - *What would the graph look like if the slope was constant?*

  - *If the slope was constant at the same rate you found between 1968 and 1974, would the y-value for the year 2020 be higher or lower than the value shown on the graph in 2020?*

  - *Do you think that would be good or bad?*

- Tell students to continue working in their groups to complete question 6, where they will choose one line segment and investigate the historical impacts of that time period.

  - Encourage students to include words like *economy, major events, president,* and *minimum wage* in their search along with their years.

- While monitoring each group, do the following:

  - Ask students why they chose the specific time frame.

  - Encourage them to use precise vocabulary to relate to the graph, line segment, and slope when appropriate.

> + Ask: *Why do you think your event may have had an impact on cost of living?*
>
> + Discuss with groups who chose nonlinear intervals why the average rate of change of their interval may or may not be a good representation of their entire time frame.
>
> + Remind students that we are looking for possible causes and effects.

- Ask students to summarize their findings on the whiteboard, a single slide, or chart paper to present to the class. Have students do a gallery review of their peers' conclusions.

- Students can have solutions written on the board, on a sheet of paper, or on a Google slide. Give students the opportunity to see their peers' work. They may write questions/comments on their own sheet of paper or directly on their peers' work with a sticky note.

- Allow students to discuss the questions and comments from their peers as they prepare to present their findings to the whole class.

SUMMARY (10 MINUTES)

- Each group will give a short presentation of their findings to the class. Presentations should include the following:

  + The time frame of the data their group chose

  + The slope of the line segment between the dates they chose

  + Their method for finding the slope

  + Any conclusions they came to

- As groups present, encourage students to record their observations on Worksheet 1. Encourage them to

  + record the slope and time frame of each group;

  + notice the difference in the representations and methods of finding slope; and

  + take notes on what is surprising from peers' presentations, and aspects they agree or disagree with.

- Pose the following questions to assess student understanding:

  + *What does a steeper slope represent?*

  + *What about a 0 slope?*

- Distribute or post the prompt for the Lesson 1 Exit Slip and have students record their answers for question 1. Use the following questions to have a brief discussion:

  + *Does a higher salary necessarily indicate more wealth?*

  + *How does cost of living affect a person/family?*

- Have students complete question 2 and then use the following questions to facilitate a discussion of the methods used to organize the slopes:

  + *What does this tell us about the cost of living over time?*

  + *How does a steeper slope or less steep slope affect daily living (if Big Macs are more/less affordable)?*

  + *What does that mean for available spending for other goods?*

  + *What can we conclude about the cost of living over time?*

Lesson 2 Facilitation

Federal Minimum Wage

Students will use the Desmos graphing calculator to construct a scatter plot and line of best fit of federal minimum wage over the same time frame as Lesson 1. Students will compare the average rate of change of minimum wage and cost of living to understand what the slope of each line (minimum wage and cost of living) represents and how the steepness of each line affects affordability.

LAUNCH AND EXPLORE (50 MINUTES)

- Distribute Worksheet 2 (*Federal Minimum Wage*), and have students work in the same groups as Lesson 1 to complete the task. Remind them to use the same interval they studied for cost of living to investigate the federal minimum wage.

- Monitor students' progress throughout the activity, encouraging the appropriate use of vocabulary (linear/nonlinear association, outlier, positive/negative correlation) and connections to the context to analyze their graphs.

- Use the task questions and the following prompts to further probe students' thinking about minimum wage and cost of living and regroup as a class to review as needed.

Question 3

+ Encourage students to consider large jumps in the graph.

+ Ask: *If you draw a line segment between these two points, does this line represent all the points in the interval?*

Question 6

+ As students use Desmos to find the line of best fit for their scatter plot, provide support to find the slope and y-intercept. Ask these questions:

  - *Is the slope found using Desmos the same or different from the slope you found in Worksheet 1?*

  - *Why or why not?*

Question 10

+ As students compare the slope they found for cost of living in Lesson 1 and the slope they found for minimum wage, ask these questions:

  - *Do units matter?*

  - *Can you accurately compare the two slopes?*

Question 10e

+ As students use their values from minimum wage and with the cost of living calculator, ask these questions:

  - *Do you think the minimum wage values are fair?*

  - *Why or why not?*

  - *Encourage students to think about how even a few cents or dollars each month can accumulate over time.*

+ Consider the budget of a typical family, and ask these questions:

  - *What are the necessities?*

  - *What should a family be able to afford on a full-time job?*

  - *What is fair?*

SUMMARIZE (5 MINUTES)

• Distribute or post the prompt for the Lesson 2 Exit Slip and have students record their answers.

Lesson 3 Facilitation

Reflection

Students will create and present an argument about when they think Big Macs were more affordable using data from this unit to support their claim. The class will summarize how slope and lines of best fit can be used to understand the relationship between two variables. Students will consider how minimum wage can affect quality of life and whether the current federal minimum wage is fair.

LAUNCH (5 MINUTES)

- Have students work in the same groups as Lessons 1 and 2 to prepare a presentation on when they think Big Macs are more affordable.

- Remind students that their presentation should include key vocabulary about their scatter plot, an analysis of slope, and historical relevance.

EXPLORE (20 MINUTES)

- Monitor students as they work in their groups to complete their presentations.

- Use purposeful selection, sequencing, and connecting techniques to determine the order for groups to present.

SUMMARIZE (30 MINUTES)

- Students will need one copy of Worksheet 3 (*Reflection*) per group whose presentation they will observe. While each group presents their findings, prompt each student observing to use Worksheet 3 to record their notes on what they agree and disagree with, what surprised them, and whether the presentation made them reconsider their position.

- Ask students to brainstorm possible counterarguments to their presentations. How could they use mathematics to justify their position?

- Encourage students to ask their peers questions at the end of their presentations.

- If students use the same piece of information to support different claims, ask these questions:

  + *What additional information could be found to support one claim over the other?*

  + *What is the difference in arguments? What is in common?*

- Distribute or post the prompt for the Lesson 3 Exit Slip and have students record their answers.

- Have students briefly discuss in their groups and then have a class discussion. Use the following prompt to facilitate a closing discussion:

    + *Did this assignment change any perspectives or beliefs you held before we began this lesson?*

    + *How can you retain a sense of empathy?*

- Encourage students to do one of the options under Taking Action.

TAKING ACTION

Encourage students to take one of the following actions:

- Write a letter to their state representatives with their conclusions and justifications about affordability of goods and services over time. Persuade state representatives to support their claim about minimum wage in 2020.

- Make a poster with their findings and a call to action. Hang it up around their school or neighborhood.

- Educate others about cost of living and minimum wage.

- Speak at a city council meeting.

- Do a similar analysis for cost of their household cooking staples over time and share results with the community.

- Explore the Fight for $15 and worker strikes at fightfor15.org and around the web. Write an article for the school newspaper summarizing current events. Read some of the articles about McDonald's and make connections to what they read and the analysis they did in the previous lessons.

- Do an internet search of efforts (including collective actions and activism) to raise the minimum wage from the past and present. Ask students to consider whether there is any action in their community.

Online Resources

 Available for download at **https://qrs.ly/wbfixtr**

Exit Slips

Day 1 Exit Slip:
1. Based on your exploration and your peers' findings, what is your answer to the question. Do you think you could buy more Big Macs with your salary in 1968 or this year? Why or why not? Did your answer change?

2. **Extension:** Organize the slopes of each group into categories or a list. Why did you choose this organization method? What does that tell us about cost of living over time?

Day 2 Exit Slip:
3. How could you justify your answer to the question, "Do you think you could buy more Big Macs with your salary in 1968 or this year?" What math can you use?

Day 3 Conclusion and Explanation:
4. When do you think people could buy the most Big Macs, in 1968 or now? Why? Did you change your mind since Day 1, when you saw the TikTok video?

5. Life is about more than just Big Macs. What about meals cooked at home? How does minimum wage affect quality of life?

Teacher Resource 1: Exit Slips

Minimum Wage and Cost of Living

Part 1: Minimum Wage
1. A Look at Minimum Wage

    a. After watching the TikTok Video, list your observations:

    b. Look at the Federal Minimum Wage graph below and list your observations:

    c. Do you think you could buy more Big Macs with your salary in 1968 or this year? Why?

2. Federal Minimum Wage (y-axis is dollars per hour, x-axis is years since 1967)

Worksheet 1: Minimum Wage and Cost of Living

Federal Minimum Wage

1. Go to the FRED economic data website to get precise values for the Federal Minimum Wage Graph (https://bit.ly/3I9ZDxf).

   *If the graph does not load, search FRED for Federal Minimum Wage.

2. Using Desmos, construct a graph to represent Federal Minimum Wage from the same time frame that your group investigated in Lesson 1. Use a table in Desmos to create this graph.

3. Find the slope from your first point to your last point in the time frame for your group. Do you think this slope is a good representative of the entire time frame? Why or why not?

4. Construct the line of best fit by following these steps:
   - In Desmos, create a new line under your xy-chart.
   - Type in $y_1 \sim mx_1 + b$.

5. What is the equation of the line of best fit?

6. What is the slope? Interpret what the slope represents. Is this the same slope you found in #3? Is it similar?

7. Does your graph have linear association? If not, choose a segment that has linear association and repeat.

8. Does your graph have outliers? If so, which point(s)?

9. Does your line of best fit have positive or negative correlation? What does this mean in the context of your graph?

Worksheet 2: Federal Minimum Wage

Reflection

When do you think people could buy the most Big Macs? Prepare your argument to present to class. Support your argument with math. Be prepared to debate!

*You may want to talk about the slope, your state's minimum wage, cost of living, your line of best fit, the historical connections, or anything else that supports your claim.

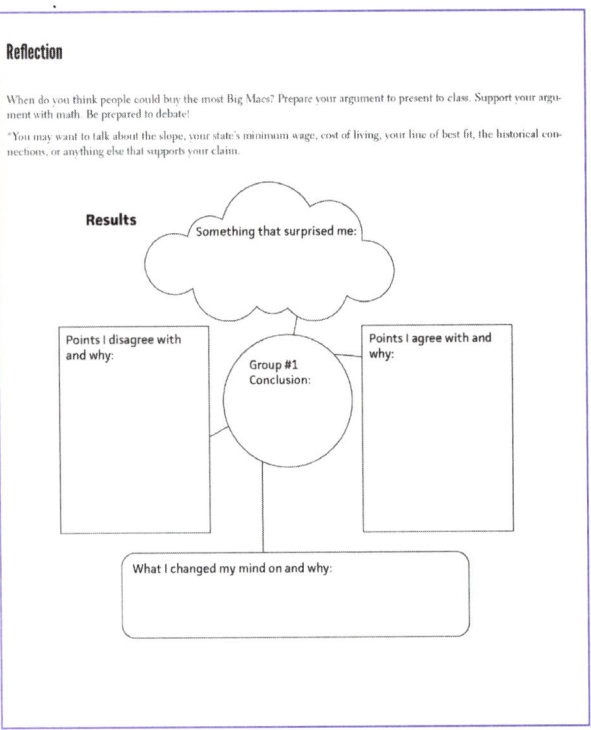

Worksheet 3: Reflection

LESSON 5.3 LISTEN TO GLSEN

Bryan Meyer and John W. Staley

Gender and Sexual Identity

In this lesson sequence, students are introduced to the GLSEN School Climate Report. The report details, among many other things, the ways that students who identify as LGBTQ+ are being mistreated in schools. Students organize data from this report into matrix form and use that to calculate the number of students in their own school who may be experiencing these forms of harassment and assault. They analyze their school's efforts to support students who identify as LGBTQ+ and identify ways they can take action to better support their classmates.

Deep and Rich Mathematics

This lesson is designed to introduce students to the use of matrices to organize large or complicated quantities of data. Students sift through raw data to organize a matrix, and they learn that there is more than one correct way to construct a matrix from a data set. Students also reinvent a method for multiplying two (or more) matrices together.

About the Lesson

The lesson is a launch–explore–summarize instructional model and is intended to take approximately 180 minutes to complete across three class periods.

Lesson 1: Students engage in sifting through lots of data in order to construct a matrix. They think about which data points suit themselves to a matrix organization and which data points do not. They also consider whether there are multiple ways to construct a matrix from a given data set.

Lesson 2: Students use two matrices in order to calculate the number of students in each grade level in their own school that are experiencing various forms of mistreatment. They work from their own intuition and common sense to calculate these values and then organize their results into a new matrix.

Lesson 3: Students learn about resources and actions they can take to better support students who identify as LGBTQ+.

Note: Taking Action is included as Lesson 3.

RESOURCES AND MATERIALS

- Task Card 1, *Listen to GLSEN* (2 copies per group)

- Task Card 2, *Closer to Home* (2 copies per group)

- Student Homework Resource 1, *Excerpt From* Think B4 You Speak (1 copy per student)

- Student Resource 2, *Taking Action* (2 copies per group)

- GLSEN 2017 National School Climate Survey Executive Summary (2 copies per group), bit.ly/2m08snE

Lesson 1 Facilitation

Listen to GLSEN

LAUNCH (20 MINUTES)

- Begin by having students journal about the following questions (they can select one). Let students know that they won't be required to share their writing.

  + *Think of a time that you were bullied or harassed. How did that make you feel?*

  + *Why do you think some people engage in bullying or harassment?*

  This is primarily an empathy-building exercise. Ask students if anyone is comfortable sharing something they wrote, but don't require it.

- Introduce the GLSEN 2017 National School Climate Survey Executive Summary. Let students know that GLSEN stands for the Gay, Lesbian,

Educators should be mindful that students may need space and time during the discussion, including the ability to leave class. Additional resources about best practices for supporting students who identify as LGBTQ+ are available in the resource guide from Teaching Tolerance: bit.ly/2kUBdSo.

and Straight Education Network. It is an organization committed to improving the experience of students who identify as LGBTQ+ in K-12 schools. The report students will look at is a collection of data about the experiences of these students in schools across the nation.

- Give students access to pp. 4-6 of the GLSEN Executive Summary ("Hostile School Climate" and "Effects of Hostile School Climate").

- Have students individually read and then discuss at their tables:

  + *What statistics stand out to you?*

  + *What questions does this bring up for you?* Record some of their noticings and (especially) their wonderings for reference later on.

- If it doesn't come up, ask:

  + *What terms were included in the report that you are unsure of the meanings of?*

- Tell students that, as time permits, they can discuss these terms in their groups and look up any definitions they are unsure of. There will also be a short homework reading that centers on the terms.

EXPLORE (20 MINUTES)

- Distribute two copies of Task Card 1, *Listen to GLSEN,* per group and have the group select a "facilitator" who will read each part of the task card and keep the group focused. Have student groups begin work together on the task.

- As students work, listen for conversations about and prompt with the following questions:

  1. *Which data from this report are suited to a matrix organization? Which data do not really seem suited to organization with a matrix?*

  2. *Our matrices look different. Can they both be correct? How would we know?*

  3. *What additional data would we need to calculate this? Why these data? What would we do with the data once we had them?*

- You could choose to have students make a poster of their work for Question 1 and/or Question 2.

If some groups seem finished, encourage them to discuss the definitions of unfamiliar terms and to look up ones that they don't know or would like to verify.

SUMMARIZE (20 MINUTES)

- Begin by having some student groups share their matrices from Question 1. In particular, center the conversation on the following:

  + *Which data were suited to matrices? Which data were more difficult to put into matrix form?*

- The intent of Question 2 is for students to see that there are multiple ways to construct a matrix from a given data set.

- Have one or two groups share two different versions of a matrix from Question 2, and ask students the following questions:

  + *Are these both correct? How do you know?*

  + *What does the 59.5 percent in this location mean? (You can do the same for the second matrix. Does the meaning change in this second matrix?)*

- After this discussion, let students know that even though there are multiple correct versions of a matrix for these data, they will use this one from here on out (you will use this in Day 2). Make a poster of this matrix and display it in your room somewhere for the remainder of the lesson sequence.

National Percentages Matrix

$$
\begin{array}{c}
 & \begin{array}{cccc} unsafe & verbal & physical & assault \end{array} \\
\begin{array}{c} sexual\ orient. \\ gender \end{array} & \left[\begin{array}{cccc} 59.5 & 70.1 & 28.9 & 12.4 \\ 44.6 & 59.1 & 24.4 & 11.2 \end{array} \right]
\end{array}
$$

- Finally, have some discussion about Question 3. This question foreshadows the task for the next lesson and is just intended to get students thinking. It is *not* important that they fully resolve this question. Focus the discussion on the following questions:

  + *What information would we need?*

  + *Why that information?*

  + *What would we do with that information if we had it?*

- For homework, ask students to read the excerpt from *Think B4 You Speak.*

Lesson 2 Facilitation

Closer to Home

- Debrief reading from the homework by asking students to answer these questions with a partner:

    + *What is one new thing/term that you learned from the reading?*

    + *What is one question you still have?*

LAUNCH (5 MINUTES)

- Re-engage students in the National Percentages Matrix from the previous day by asking these questions:

    + *What do you remember about what the data in the matrix represent?*

    + *What do these two numbers mean, specifically?* (Choose any two values from the same column in order to distinguish that one represents the percentage of students nationwide who are experiencing that based on sexual orientation, and the other is the percentage of students nationwide who are experiencing that based on gender expression.)

- Post on the board the total number of ninth, tenth, eleventh, and twelfth graders in your school (this needs to be visible for everyone for the task). Then, ask the following question:

    + *How many students in our own school do you think are experiencing this type of mistreatment?*

EXPLORE (20 MINUTES)

- Distribute Task Card 2, *Closer to Home*. Even though students will be working in groups of four, it may be a good idea to give each student a copy so that everyone has access to the images on the card.

- Have one person in each group read through the whole task card and then ask the group members to restate the task in their own words. You might choose to have some brief whole-class discussion prior to setting groups off to work.

- Begin by assigning a few minutes of "private reasoning time" for students. This should help each group member to begin thinking for themselves before they collaborate.

- In Question 1, the intent is that students will take the data from their own school (by grade level) and multiply by 8 percent (for sexual orientation) and 1 percent (for gender expression) to find approximate numbers of students at their own school who identify in these ways. For instance, if a school had 625 ninth graders, 750 tenth graders, 800 eleventh graders, and 500 twelfth graders, their calculations would look something like this:

My School Matrix

$$
\begin{array}{c}
& \begin{array}{cc} \textit{sexual orient.} & \textit{gender} \end{array} \\
\begin{array}{c} \textit{Ninth grade} \\ \textit{Tenth grade} \\ \textit{Eleventh grade} \\ \textit{Twelfth grade} \end{array}
&
\left[
\begin{array}{cc}
625(.08) & 625(.01) \\
750(.08) & 750(.01) \\
800(.08) & 800(.01) \\
500(.08) & 500(.01)
\end{array}
\right]
\end{array}
$$

Resulting in this "My School Matrix" →

My School Matrix

$$
\begin{array}{c}
& \begin{array}{cc} \textit{sexual orient.} & \textit{gender} \end{array} \\
\begin{array}{c} \textit{Ninth grade} \\ \textit{Tenth grade} \\ \textit{Eleventh grade} \\ \textit{Twelfth grade} \end{array}
&
\left[
\begin{array}{cc}
50 & 6.25 \\
60 & 7.5 \\
64 & 8 \\
40 & 5
\end{array}
\right]
\end{array}
$$

- As groups work on the task, you may want to briefly pause the class at times to

  + remind students that the numbers in the National Percentages Matrix are percentages, so they will need to convert to decimal to multiply (or do some other calculation that makes sense to them).

  + have a group share their calculations for Question 2, Part a. Then, have groups continue working on the rest.

SUMMARIZE (40 MINUTES)

- The intent in this section is to work from students' own intuition and natural sensemaking in order to build a definition for matrix multiplication. In a sense, students reinvent this procedure based on the repeated calculations that they do (especially in Question 2).

- To begin, you might have a student group present their calculations from Question 2a. *At this point, students should be using the My School Matrix based on data from your own school.* For the purposes of illustration, hypothetical numbers are presented here (yours will likely be different). If the My School Matrix were filled with the numbers below, that calculation would look something like this:

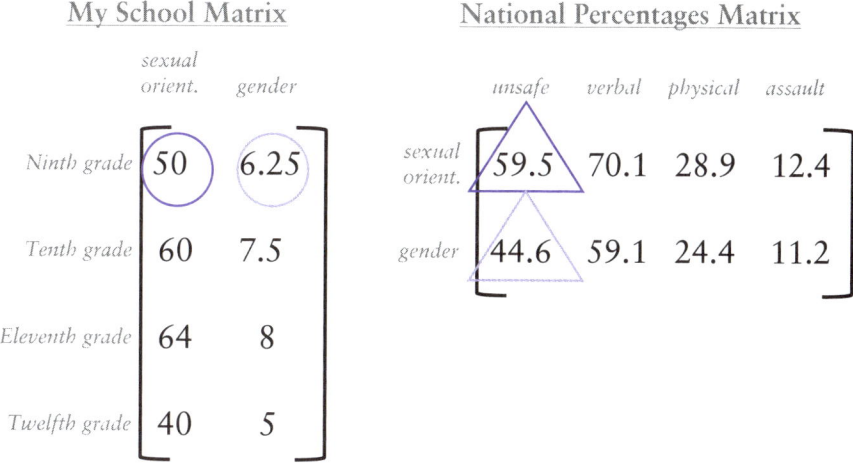

$$50\,(.595) + 6.25\,(.446) = 32.5375$$

- Ask the following questions:

    + *Why did it make sense to multiply 50 × .595? And 6.25 × .446?*

    + *What does our answer (32.5375) mean in this case?*

- Introduce the solution matrix if it hasn't already come up:

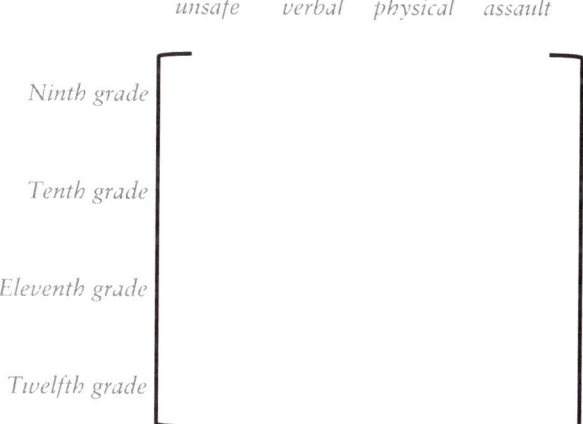

- It will probably be useful for students to present their calculations for at least two of the parts (i-iv). This will allow the class to see the repeated reasoning and calculation. Ask the class:

    + *What is the process you are doing over and over again?*

- Be sure to generalize here that *to get each entry of the product of two matrices, multiply a row in the first matrix times a column in the second matrix.* Tell students that this is the process of matrix multiplication. Select another entry of the solution matrix (eleventh graders who are experiencing assault, for instance) and ask students how they would get that entry.

- At this point, introduce the following alternative My School Matrix. The purpose for introducing this is to help students begin to generalize about when a matrix can and can't be multiplied together. Seeing this nonexample will help them with that.

<u>My School Matrix</u> $\times$ <u>National Percentages Matrix</u>

$$
\begin{array}{c}
 \\
\text{Ninth grade} \\
\text{Tenth grade} \\
\text{Eleventh grade} \\
\text{Twelfth grade}
\end{array}
\begin{array}{ccc}
\textit{sexual} & & \\
\textit{orient.} & \textit{gender} & \textit{questioning} \\
\end{array}
\begin{bmatrix}
50 & 8 & 1 \\
60 & 10 & 0 \\
75 & 12 & 3 \\
40 & 6 & 8
\end{bmatrix}
$$

$$
\begin{array}{c}
 \\
\textit{sexual} \\
\textit{orient.} \\
\textit{gender}
\end{array}
\begin{array}{cccc}
\textit{unsafe} & \textit{verbal} & \textit{physical} & \textit{assault} \\
\end{array}
\begin{bmatrix}
59.5 & 70.1 & 28.9 & 12.4 \\
44.6 & 59.1 & 24.4 & 11.2
\end{bmatrix}
$$

- Ask the following questions:

  + *Can we multiply these two matrices in the same way? Why or why not?*

  + *What needs to be true in order to multiply two matrices?*

  The key here is that the length of the rows in the first matrix needs to equal the height of the columns in the second matrix.

- Finally, if students haven't already done so, have them finish finding the values for all entries of the solution matrix. While this may take some time, the numbers represent the approximate number of students at their school who are being mistreated. So, it is important to finish this task. You could divide the work among the class to save time. Ask students this question:

  + *What are your reactions to seeing these numbers about our school?*

- Related, the numbers students get as a result of the matrix multiplication are just approximations and aren't likely to exactly match the actual numbers from your school. Ask students to consider this question:

  + *What are some reasons that these numbers may not be exact?*

  Some reasonable answers could include the following:

  - The national percentages of students who identify with non-dominant sexual orientations (8 percent) and/or gender expressions (1 percent) are approximations themselves.

- Some students who identify as LGBTQ+ may be scared to report harassment, in which case our actual numbers would be higher.

- We may have double-counted.

- Some students may identify with both nondominant sexual orientations *and* gender expressions, and thus reported harassment for both reasons.

• You do not need to spend time discussing the challenge question with the whole class. Should you have students who are interested, below is one possible way of setting up that multiplication:

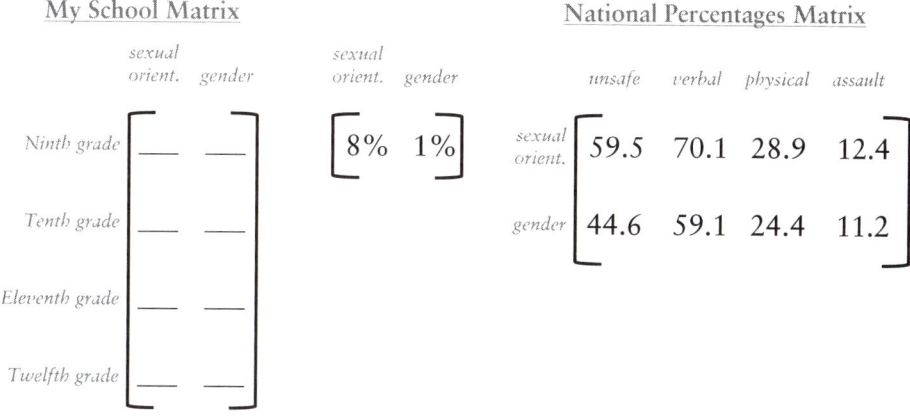

You might recognize that the multiplication of the first two matrices is another way of setting up the calculations that students already did in Question 1.

Lesson 3 Facilitation

Taking Action (Considered as a third full lesson period)

LAUNCH (10 MINUTES)

• Begin by having students journal answers to the following questions:

  + *What is our school doing to help prevent this type of mistreatment?*

  + *What is our school doing that might be contributing to this type of mistreatment?*

• Have students discuss their ideas in small groups, and then take (and record) some responses for the whole class to hear.

EXPLORE (20 MINUTES)

- Project the two graphs from Student Resource 2, *Taking Action*. (The graphs show longitudinal trends in both victimization and school supports.) Ask students these questions:

    + *What do you notice? What do you wonder?*

- They will likely comment that incidents of victimization seem to be decreasing over time. Also, school support services seem to be increasing over time. They may also comment about the connection between these two (for instance, that incidents may be decreasing *because of* the increase in services).

- These services (school policies, practices, and procedures that can help reduce these incidents on campuses) are outlined in greater detail in the GLSEN School Climate Report, in "LGBTQ-Related School Resources and Supports" on pp. 7-10. The services fall under four domains:

    + GSAs (Gay-Straight Alliances)

    + Inclusive curricular resources

    + Supportive educators

    + Inclusive and supportive school policies

- Organize a jigsaw for students to learn about these. Assign each group to read and become "experts" on one of the sections. Then, form mixed groups and have the students teach each other about the ones they didn't read.

SUMMARIZE (30 MINUTES)

- Use the following questions to facilitate a class discussion:

    + *What steps do we need to take at our school to better support our students who identify as LGBTQ+?* (List these on the board.)

    + *Which of these actions do we feel most compelled to take?*

    + *How can we organize ourselves to follow through on these actions?*

        - *Should we form small groups?*

        - *Who would take the lead?*

        - *What would a timeline for this look like?*

    + *Who else on campus would need to be involved?*

Worksheets and Teacher resources

Excerpt from Think B4 You Speak

This section adapted from the Think B4 You Speak Educator's Guide (p. 4) and GLSEN Gender Terminology, www.glsen.org

Lesbian, gay, bisexual and transgender (LGBT) teens in the U.S. experience homophobic remarks and harassment throughout the school day, creating an atmosphere where they feel disrespected, unwanted and unsafe. GLSEN's

2007 National School Climate Survey found that nearly three-quarters (73.6%) of LGBT students hear homophobic language, such as "faggot" or "dyke," and more than nine in ten (90.2%) hear the word "gay" used in a negative way frequently or often at s chool. Though many play down the impact of expressions like "that's sogay" because they have become such a common part of teens' vernacular and are often not intended to inflict harm, 83.1% of LGBT students say that hearing "gay"or "queer" used in a negative manner causes them to feel bothered or distressed.

> ### DID YOU KNOW?
> 9 out of 10 LGBT students hear the word "gay" used in a negative way often or frequently in school and nearly 3 out of 4 LGBT students report hearing their peersmake homophobic remarks, such as "dyke" or "faggot," often or frequently in school.

Studies indicate that youth who regularly experience verbal or physical harassment suffer from emotional turmoil, low self-esteem, loneliness, depression, poor academic achievement and high rates of absenteeism.[1] Research also shows that many of the bystanders to acts of harassment experience feelings of helplessness and powerlessness, and develop poor coping and problem-solving skills.[2] Clearly, homophobic and all types of harassment—and the toxic effects they produce—are whole school problems that all educators must confront.

To address this disconcerting reality, the Gay, Lesbian and Straight Education Network (GLSEN) together with The Advertising Council has created the first national multimedia public service advertising (PSA) campaign designed to address the use of anti-LGBT language among teens. The campaign aims to raise awareness among straight teens about the prevalence and consequences of anti-LGBT bias and behavior in America's schools. Ultimately, the goal is to reduce and prevent the use of homophobic language in an effort to create a more positive environment for LGBT teens. The campaign also aims to reach adults, including school personnel and parents, because their support of this message is crucial to the success ofefforts to change teens' behavior.

A Note About Language

This section adapted from the *Think B4 You Speak Educator's Guide* (p. 10) and GLSEN Gender Terminology, www.glsen. org

There are many terms that are used to describe what is commonly known as the "gay community." Since the word "gay" most often refers to homosexual men, we have chosen to use the more inclusive "LGBT," which means lesbian, gay, bisexual and transgender. Though this term may sound unfamiliar at first, we encourage you to use it consistently with [others] and to avoid reflexively using "gay" to describe the broad spectrum of sexual and gender identities. There are many people

▲ *Student*
Homework Resource 1

Taking Action

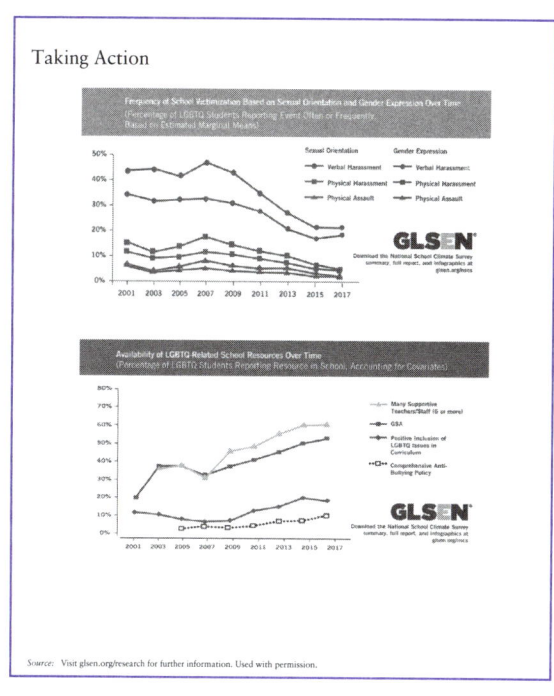

Source: Visit glsen.org/research for further information. Used with permission.

▲
Student Resource 2

Listen to GLSEN

Find some information in the provided section of the GLSEN 2017 National School Climate Survey Executive Summary (pp. 4–6) that could be organized into a matrix. Create a matrix for that information. Be sure to label your rows and columns.

1. Below is a slightly simplified version of one section of the GLSEN report. Have every member of your group organize this information into a single matrix (be sure to have everyone label their rows and columns). Then, compare your matrices and decide on one matrix for your group.

> 59.5 % of LGBTQ students felt unsafe at school because of their sexual orientation and 44.6% because of their gender expression.
>
> 70.1% of LGBTQ students experienced verbal harassment (e.g., called names or threatened) at school based on sexual orientation and 59.1% based on gender expression.
>
> 28.9% of LGBTQ students were physically harassed (e.g., pushed or shoved) in the past year based on sexual orientation and 24.4% based on gender expression.
>
> 12.4% of LGBTQ students were physically assaulted (e.g., punched, kicked, injured with a weapon) in the past year based on sexual orientation and 11.2% based on gender expression.

2. How could you use the statistics in Question 2 to find out how many students at your own school are being affected in these ways? What additional information would you need?

▲
Task Card 1

Closer to Home

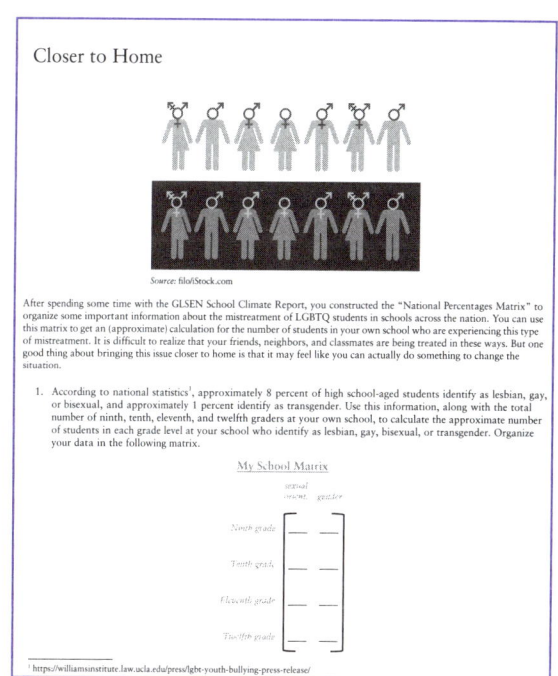

Source: filoiiStock.com

After spending some time with the GLSEN School Climate Report, you constructed the "National Percentages Matrix" to organize some important information about the mistreatment of LGBTQ students in schools across the nation. You can use this matrix to get an (approximate) calculation for the number of students in your own school who are experiencing this type of mistreatment. It is difficult to realize that your friends, neighbors, and classmates are being treated in these ways. But one good thing about bringing this issue closer to home is that it may feel like you can actually do something to change the situation.

1. According to national statistics[1], approximately 8 percent of high school-aged students identify as lesbian, gay, or bisexual, and approximately 1 percent identify as transgender. Use this information, along with the total number of ninth, tenth, eleventh, and twelfth graders at your own school, to calculate the approximate number of students in each grade level at your school who identify as lesbian, gay, bisexual, or transgender. Organize your data in the following matrix.

My School Matrix

sexual
orient. gender

Ninth grade

Tenth grade

Eleventh grade

Twelfth grade

[1] https://williamsinstitute.law.ucla.edu/press/lgbt-youth-bullying-press-release/

▲
Task Card 2

LESSON 7.5 HUMANIZING THE IMMIGRATION DEBATE

Ayse Ozturk and Stephen Lewis

Immigration

Though this lesson is explicitly centered on the Arizona Sonoran Desert, it provides an excellent opportunity for students to begin to share the lived experiences of immigrants. According to the Arizona OpenGIS Initiative for Deceased Migrants, since 2001 there were a total of 127 documented deaths of migrants crossing this desert in the United States. Lessons like this are intended to empower students to reach statistically sound conclusions while connecting to the hardships immigrants face, such as lack of food and water while trying to cross the southern US border.

Deep and Rich Mathematics

Students collect, summarize, represent, and interpret statistical data to develop understanding of immigration. Through this context, students advance understanding of data trends and how the variables are related through dot plots, histograms, and boxplots. Students work with real-world problems by using mathematics as a language for understanding, simplifying, and solving these situations in an interdisciplinary method (Bassanezi, 2002).

About the Lesson

This lesson follows a Three-Act Task model and is intended to take approximately 270 minutes to complete across three class periods.

Act 1: Students learn about issues faced by people who are trying to cross the border illegally.

Act 2: Students begin to examine and graphically represent migrant mortality data.

Act 3: Students make inferences about migrant mortality data.

RESOURCES AND MATERIALS

- Lesson agenda on PowerPoint

- Computer with internet access and projector

- 80-minute documentary available at nominal cost, "Who Is Dayani Cristal?," whoisdayanicristal.com

- Notice and Wonder worksheet

- Website for data collection, humaneborders.info

- Video, "Volunteering With Humane Borders," bit.ly/2kY5pMD

- You may be interested in another social justice math lesson (SJML) on the immigration by Turner, Varley Gutiérrez, Simic-Muller, and Díez-Palomar (2009).

This lesson was developed using the following resources: Anderson et al. (2001); Daniels and Zemelman (2004); NCTM (2000); Ohio Department of Education (2015); Rosa and Orey (2007).

- Statistics and Probability—Scatterplots, including plots over time, can reveal patterns, trends, clusters, and gaps that are useful in analyzing the association between two contextual variables. (VSD.4)

MATHEMATICAL PRACTICES
- Model with mathematics.

- Construct viable arguments and critique the reasoning of others.

Lesson 1 Facilitation

ACT 1: "WHO IS DAYANI CRISTAL?" (90 MINUTES)

Students explore the immigration debate learning more about Dayani Cristal, a Mexican immigrant wanting to cross the border illegally. This lesson is a powerful starting point for studies on US immigration, as well as domestic issues influenced by undocumented immigrants.

- Provide the following prompt for students to consider as a think-pair-share before exploring the documentary for class:

  + *Why do you think people migrate?*

- Watch the documentary showing the perspective of a young Mexican man with limited opportunities who wants to illegally cross the US/Mexico border.

- At specific points during the documentary, have students stop and reflect in writing using the Notice and Wonder worksheet.

- The following are some possible questions to ask when pausing the documentary:

  + *How does socioeconomic class affect people's choices or abilities to achieve their goals?*

  + *Why is it mostly men who leave their home and family?*

  + *Do you think migrants' expectations about life in the United States are realistic? Why or why not?*

  + *If you had a chance to talk with Dayani, what would you tell her and why?*

  + *What are the dangers during migration across the US/Mexico border?*

  + *Imagine yourself in Dayani's situation, preparing for a dangerous journey and leaving home, perhaps for the last time, but with the hope of a better future. How would you prepare yourselves?*

  + *Discuss what economic, political, cultural, historical, and geographical forces are influencing the choices that Dayani's father is making about his future. Ask students to draw from their observations in the film, current events, and/or personal experiences to support points that they make.*

- Lead a class discussion on students' notices and wonders about the documentary.

- Use the following questions to promote connections to the next lesson as a summary to Lesson 1.

  + *How do the wages in Central America compare to the United States? How might this serve as a factor for migration?*

  + *What variables might impact one's choice to migrate?* (List them)

+ *What would be essentials to take on a journey across the Sonoran Desert? How much of each item would you require to optimize chances of survival?*

+ *How does the temperature in the desert compare to where you live? How does this temperature impact water consumption necessity?*

- Emphasize the difference between statistical and nonstatistical questions.

Lesson 2 Facilitation

ACT 2: PROBLEM CREATION AND LOOKING AT IMMIGRATION DATA (90 MINUTES)

The goal of this portion of the lesson is to synthesize the highly contextualized information from earlier. Students begin to develop the ability to examine data and represent it in multiple ways.

- Allow time for students to start up digital devices to explore a website.

- Introduce Arizona OpenGIS Initiative for Deceased Migrants, humaneborders.info, and show how to do research on migrant mortality in different categories (e.g., year, cause of death, county of death, gender).

- Place students into groups of four.

- Have each group create a statistical question that requires research on migrants' deaths. For example, while one of the groups works on the question "How many male migrants die because of exposure in a given ten-year period?," another other group might work on the question "In which region of Arizona desert do people mostly die?"

- Have students work on any questions they are wondering about.

- Have students, as a group, decide together what to search for on the website. They should take notes and have a group discussion.

- During the group study time, walk around and check each group's work and offer help or guidance as needed.

- Have groups create a data table with ten cases under their research question. For example, one group might create a table to document the number of/reasons for female deaths in the past five years. Some students may be confused at this point because there are seven variables on the website and it's easy to get lost when searching information.

- Ensure students understand experimentation and the need to keep one variable changeable and the others constant. For example, if a group looks for how many people have died over the past five years, the variables (i.e., land county, cause of death, gender) should be filtered as "any" and the year of death variable should be specifically selected (2016, 2017, etc.).

- Monitor each group's work to ensure students are accurately collecting data.

- While circling the room during the group work part of this lesson, ask questions that pertain to each group's particular method and thinking, such as the following:

  + *Describe to me how you decided your research question.*

  + *What variables are important to be able to create your data table? Why? How are they different?*

  + *How could you use a table to show me any patterns you figured out during your investigation? What is the best way to display your data (data table, graph, list, etc.)? Why would this method be optimal as compared to others?*

  + *What patterns or trends exist in the data when examining these variables: males to females; cause of death; regions with significant mortality?*

  + *How can deaths identified as "unknown" or "skeletal remains" be distributed based on the known causes and frequencies of death?*

- To summarize the lesson, have each group present their research question, data collection method, and table of data.

- Facilitate discourse around patterns in their data set and sound statistical conclusions. After each presentation, ask the following questions:

  + *What inferences can you make from the ten cases you examined?*

  + *Did you notice any pattern in your data table?*

- To conclude the class, ask students:

  + *What would be another way to represent your data?*

  + *How will you plan to monitor and explain the variation in your data?*

  + *Can we make a graph for the data you collected? What would it look like?*

Lesson 3 Facilitation

ACT 3: DATA REPRESENTATION AND MAKING INFERENCES (90 MINUTES)

Students interpret and compare data from Act 2. They present graphs to the class and make inferences based on the data they collected in the previous lesson, taking into account the shape, center, spread, and any potential unordinary data points.

- Discuss different types of representations (bar graphs, histograms, line graphs, scatterplots, etc.) as a way to highlight how different representations can showcase different aspects of a context.

- Have students return to their original groups and create graphical representations of their data tables as a way of illustrating relationships between collected variables from the Humane Borders website.

- Distribute large chart papers and colored markers to each group.

- Have student groups share their graphical representations with the class outlining key findings or interpretations or trends the data reveal. In summarizing categorical data, it might be beneficial to have students begin with two categories for each variable and represent them in a two-way table with the two values of one variable defining the rows and the two values of the other variable defining the columns.

- Use the following questions as a way to facilitate discussion among the students:

  + *What are the results for the data you collected?*

  + *What's the story behind your graph?*

  + *Is there another way to present your data?*

  + *How are bar graphs different from histograms?*

  + *How does your graphical display share variability and center?*

  + *What are your assumptions on the causes of your conclusion? Why?*

  + *Based on your data representation's center, spread, and shape, what do you expect to happen in forthcoming years?*

- As students collect data from the Humane Borders website, the following trends will emerge:

  + While other causes of death are present within the data, the highest proportion of identified migrant deaths is due to exposure caused by water scarcity.

+ A large number of deaths are listed as "unidentified" or "skeletal remains."

- Discuss how to allocate the unidentified deaths into the major categories listed by Humane Borders.

- Discuss how the trends exhibited within the past five years can better predict migrant death in the years to come. This creates a link to the next lesson, where students are asked to consider how to minimize the number of deaths due to exposure by placing water stations across the desert. Consider the following prompts during discourse:

  + *How can we use our data to predict future occurrences of mortality in the Sonoran Desert?*

  + *In what ways could mortality be reduced? What variables would impact this reduction?*

  + *How much water does one person need to consume in a trip across the Sonoran Desert?*

  + *How much food does one need to consume?*

  + *How long would you predict a trip across the Sonoran Desert would take? What route would be optimal for this journey?*

TAKING ACTION

The goal of this final portion of the lesson is to help students determine how many water stations consisting of 55-gallon drums of water would be sufficient to minimize migrants' deaths due to dehydration.

- Show the video "Volunteering With Humane Borders," bit.ly/2kY5pMD.

- Pose the question to students, *How many water stations are needed to minimize migrant mortality from dehydration?*

- Ensure students consider these variables:

  + Amount of water needed to survive a day in the desert

  + Amount of time one could reasonably walk during a day in the desert

  + Distance across the desert between major cities or landmarks (e.g., Nogales and Tucson or Tohono O'odham Nation reservation and Phoenix)

  + Reasonable distances between water stations assuming migrants are carrying gallon jugs, portable water bottles, or nothing

- Across this discussion, you should support students in the following:

  + Making assumptions to control for variables

  + Stripping away less relevant variables to establish initial values

  + Correlating water needed to anticipated number of migrant deaths based on mortality data

  + Placing water stations based on mortality traffic through the desert

  + Considering how much it would cost to supply this water over an extended period of time (one month, three months, one year)

  + Raising capital to support Humane Borders in their effort to reduce migrant mortality

- Use the following prompts to facilitate student learning:

  + *Given that the primary cause of migrant mortality across the Sonoran Desert is exposure to sun and dehydration, how many water stations would be sufficient to minimize migrant mortality?*

  + *How much money would be needed in order to purchase these water stations consisting of 55-gallon drums?*

  + *What is the predicted impact of placement of these water stations? How do you know?*

  + *Where would be optimal locations for placing these water stations?*

- Encourage students to contact Humane Borders and report their findings to determine whether Humane Borders has sufficient water stations in place.

- Have students present or share their estimates and rationale with considered contextual factors.

- Have students critique each other's reasoning and compare estimates across groups.

- Come to a class consensus on a sufficient number of water stations.

- Connect students with Humane Borders or other volunteer groups to help promote their mission of reducing migrant mortality.

- Encourage students to collect and send money to volunteer groups engaged in reducing immigrant mortality.

- Support presentations to the school or local community to increase awareness about this global issue as well as to inform others about what they learned about migration during their research.

Worksheets and Teacher resources

online resources ► These downloadable resources can be found online at **https://qrs.ly/wbfixtr**

◄ *Slide Deck*

HUMANIZING THE IMMIGRATION DEBATE

Make a list that includes your notices/wonders about the documentary.

What did you NOTICE on this documentary?	What do you WONDER?

► *Notice and Wonder Worksheet*

REFERENCES

Aguilar, E. (2021). *The art of coaching workbook: Tools to make every conversion count.* Jossey-Bass.

Aguirre, J. M., & Civil, M. (Eds.). (2016). Mathematics education: Through the lens of social justice [Special issue]. *Teaching for Excellence and Equity in Mathematics, 7*(1). https://www.todos-math.org/assets/documents/TEEM/teem7_final1.pdf

Aguirre, J., Marfield-Ingram, K., & Martin, D. B. (2013). *The impact of identity in K-8 mathematics teaching: Rethinking equity-based practices.* National Council of Teachers of Mathematics.

Ambroso, E., Dunn, L., & Fox, P. (2021). *Research in brief: Engaging and empowering diverse and underserved families in schools.* https://ies.ed.gov/ncee/edlabs/regions/west/relwestFiles/pdf/Family_Engagement_and_Empowerment_Brief_Final_Clean_ADA_Final.pdf

Bartell, T. G. (2012). Is this teaching mathematics for social justice? Teachers' conceptions of mathematics classrooms for social justice. In A. A. Wager & D. W. Stinson (Eds.), *Teaching mathematics for social justice conversations with educators* (pp. 113-125). National Council of Teachers of Mathematics.

Bartell, T. G., Yeh, C., Felton-Koestler, M. D., & Berry, R. Q., III. (2022). *Upper elementary mathematics lessons to explore, understand, and respond to social injustice.* Corwin.

Benjamin Banneker Association. (2017). *Implementing a social justice curriculum: Practices to support the participation and success of African-American students in mathematics* [Position Statement]. https://bbamath.org/wp-content/uploads/2017/11/BBA-Social-Justice-Position-Paper_Final.pdf

Berry, R. Q., Conway, B. M., Lawler, B. R., & Staley, J. W. (2020). *High school mathematics lessons to explore, understand, and respond to social injustice.* Corwin.

Boaler, J. (1993). The role of contexts in the mathematics classroom: Do they make mathematics more "real"? *For the Learning of Mathematics, 13*(2), 12-17.

Boston, M. (2012). Assessing instructional quality in mathematics. *The Elementary School Journal, 113*(1), 76-104.

Carnoy, M., & García, E. (2017). Five key trends in U.S. student performance. *Economic Policy Institute.* https://files.epi.org/pdf/113217.pdf

CASEL. (n.d.). *Advancing social and emotional learning.* https://casel.org

Center for Teaching Innovation. (2024). *Collaborative learning.* https://teaching.cornell.edu/teaching-resources/active-collaborative-learning/collaborative-learning

Chardin, M., & Novak, K. R. (2020). *Equity by design: Delivering on the power and promise of UDL.* Corwin.

Childs, K., & Davis, T. (2023, July 11). *5 tips to becoming an equity-focused educator* [Conference presentation]. HIVE 2023 Conference, Atlanta, GA, United States.

Choi, J. (2008). Unlearning colorblind ideologies in education class. *The Journal of Educational Foundations, 22*(3/4), 53-71.

Cintron, S. M., Wadlington, D., & Chen Feng, A. (2021). *A pathway to equitable math instruction: Dismantling racism in mathematics instruction (Stride 1).* https://equitablemath.org/wp-content/uploads/sites/2/2020/11/1_STRIDE1.pdf

Colorado Department of Education. (n.d.). *Culturally responsive practices and achieving* instructional equity [Online course]. https://sitesed.cde.state.co.us/mod/book/view.php?id=8030&chapterid=8126

Conway, B. M., Id-Deen, L., Raygoza, M. C., Ruiz, A., Staley, J. W., & Thanheiser, E. (2022). *Middle school mathematics lessons to explore, understand, and respond to social injustice*. Corwin.

DiAngelo, R. (2011). White fragility. *International Journal of Critical Pedagogy, 3*(3), 54-70.

DuFour, R., DuFour, R., Eaker, R., Many, T. W., & Mattos, M. (2016). *Learning by doing: A handbook for professional learning communities at work*. Solution Tree Press.

Eberhardt, J. L. (2019). *Biased: Uncovering the hidden prejudice that shapes what we see, think, and do*. Penguin Books.

Gay, G. (2002). Preparing for culturally responsive teaching. *Journal of Teacher Education, 53*(2), 106-116.

Gay, G. (2018). *Culturally responsive teaching: Theory, research, and practice*. Teachers College Press.

Gutiérrez, R. (2012). Context matters: How should we conceptualize equity in mathematics education? In B. Herbel-Eisenmann, J. Choppin, D. Wagner, & D. Pimm (Eds.), *Equity in discourse for mathematics education: Theories, practices, and policies* (pp. 17-33). Springer.

Gutstein, E. (2003). Teaching and learning mathematics for social justice in an urban, Latino school. *Journal for Research in Mathematics Education, 34*(1), 37-73.

Joseph, G. G. (2011). *The crest of the peacock: Non-European roots of mathematics*. Princeton University Press.

Key Differences. (2024). *Difference between collaborative learning and cooperative learning*. https://keydifferences.com/difference-between-collaborative-learning-and-cooperative-learning.html

Kilpatrick, J., Swafford, J., & Findell, B. (Eds.). (2001). *Adding it up: Helping children learn mathematics*. National Research Council, Mathematics Learning Study Committee. National Academy Press.

Koestler, C., Ward, J., Zavala, M., & Bartell, T. G. (2022). *Early elementary mathematics lessons to explore, understand, and respond to social injustice*. Corwin.

Ladson-Billings, G. (1995). But that's just good teaching! The case for culturally relevant pedagogy. *Theory Into Practice, 34*(3), 159-165.

Ladson-Billings, G. (2009). *The dreamkeepers: Successful teachers of African American children* (2nd ed.). Jossey-Bass.

Ladson-Billings, G. (2021). I'm here for the hard re-set: Post pandemic pedagogy to preserve our culture. *Equity and Excellence in Education, 54*(1), 68-78.

Learning for Justice. (2016). *Social justice standards: The teaching tolerance anti-bias framework*. https://www.learningforjustice.org/sites/default/files/2017-06/TT_Social_Justice_Standards_0.pdf

Liljedahl, P. (2020). *Building thinking classrooms in mathematics grades K-12: 14 teaching practices for enhancing learning*. Corwin.

McIntosh, P. (1998). White privilege: Unpacking the invisible knapsack. In M. McGoldrick (Ed.), *Re-visioning family therapy: Race, culture, and gender in clinical practice* (pp. 147-152). Guilford Press.

Merriam-Webster. (n.d.). Folx. In *Merriam-Webster.com dictionary*. https://www.merriam-webster.com/dictionary/folx

National Center for Education Statistics. (2022). *NAEP mathematics assessment*. https://www.nationsreportcard.gov/highlights/mathematics/2022/

National Council of Supervisors of Mathematics, & TODOS. (2016). *Mathematics education through the lens of social justice: Acknowledgment, actions, and accountability*. https://www.todos-math.org/assets/docs2016/2016Enews/3.pospaper16_wtodos_8pp.pdf

National Council of Teachers of Mathematics. (2000). *Principles and standards for school mathematics*.

National Council of Teachers of Mathematics. (2014). *Principles to actions: Ensuring mathematical success for all*.

National Council of Teachers of Mathematics. (2020). *Catalyzing change in early childhood and elementary mathematics: Initiating critical conversations*.

National Council of Teachers of Mathematics. (2023). *Equitable integration of technology for mathematics learning* [Position statement]. https://www.nctm.org/Standards-and-Positions/Position-Statements/Equitable-Integration-of-Technology-for-Mathematics-Learning/

National Governors Association Center for Best Practices & Council of Chief State School Officers (2010). *Common core state standards for mathematics*. https://corestandards.org/wp-content/uploads/2023/09/Math_Standards1.pdf

OECD. (2019). *The learning compass 2030.* https://www.oecd.org/education/2030-project/

Okun, T. (1999). *White supremacy culture.* https://www.whitesupremacyculture.info/uploads/4/3/5/7/43579015/okun_-_white_sup_culture.pdf

Osta, K., & Vasquez, H. (n.d.). Implicit bias and structural racialization. *National Equity Project.* https://www.nationalequityproject.org/frameworks/implicit-bias-structural-racialization

P21. (2019). *Framework for 21st century learning.* Battelle for Kids. https://www.battelleforkids.org/networks/p21/frameworks-resources

Rabbitt, M. P., Hales, L. J., Burke, M. P., & Coleman-Jensen, A. (2023). *Household food security in the United States in 2022* (Report No. ERR-325). U.S. Department of Agriculture, Economic Research Service. https://doi.org/10.32747/2023.8134351.ers

Smith, M. S., & Stein, M. K. (1998). Selecting and creating mathematical tasks: From research to practice. *Mathematics Teaching in the Middle School, 3,* 344–350.

Stein, M., Smith, M., Henningsen, M., & Silver, E. (2000). *Implementing standards-based mathematics instruction: A casebook for professional development.* Teacher College.

Teaching Tolerance. (2016). *Social justice standards: The Teaching Tolerance anti-bias framework.* Southern Poverty Law Center.

UNESCO-UNEVOC International Centre. (n.d.). Cooperative learning. In *TVETipedia Glossary.* https://unevoc.unesco.org/home/TVETipedia+Glossary/lang=en/show=term/term=Cooperative+learning

United Nations. (2020). *On World Social Justice Day, the UN labour agency says "put people and planet first."* https://news.un.org/en/story/2020/02/1057811

Wager, A., & Stinson, D. (Eds.). (2012). *Teaching mathematics for social justice: Conversations with educators.* National Council of Teachers of Mathematics.

Welsing, F. C. (2004). *The ISIS papers: The keys to the colors.* Third World Press.

INDEX

**JENNIFER M. BAY-WILLIAMS,
JOHN J. SANGIOVANNI,
ROSALBA SERRANO,
SHERRI MARTINIE,
JENNIFER SUH, C. DAVID WALTERS**

Because fluency is so much more
than basic facts and algorithms.
Grades K–8

**ROBERT Q. BERRY III, BASIL M. CONWAY IV,
BRIAN R. LAWLER, JOHN W. STALEY,
COURTNEY KOESTLER, JENNIFER WARD,
MARIA DEL ROSARIO ZAVALA,
TONYA GAU BARTELL, CATHERY YEH,
MATHEW FELTON-KOESTLER,
LATEEFAH ID-DEEN,
MARY CANDACE RAYGOZA,
AMANDA RUIZ, EVA THANHEISER**

Learn to plan instruction that engages
students in mathematics explorations
through age-appropriate and culturally
relevant social justice topics.
**Early Elementary, Upper Elementary,
Middle School, High School**

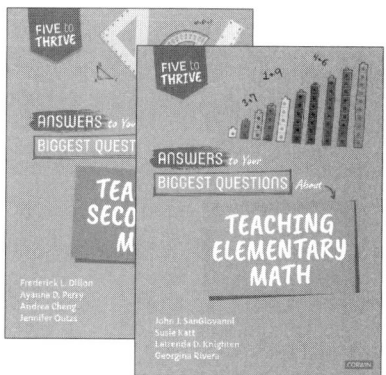

**JOHN J. SANGIOVANNI, SUSIE KATT,
LATRENDA D. KNIGHTEN,
GEORGINA RIVERA,
FREDERICK L. DILLON,
AYANNA D. PERRY,
ANDREA CHENG, JENNIFER OUTZS**

Actionable answers to your most
pressing questions about teaching
elementary and secondary math.
Elementary, Secondary

**SARA DELANO MOORE,
KIMBERLY RIMBEY**

A journey toward making
manipulatives meaningful.
Grades K–3, 4–8

A Sage Company

CORWIN HAS ONE MISSION: to enhance education through intentional professional learning.

We build long-term relationships with our authors, educators, clients, and associations who partner with us to develop and continuously improve the best evidence-based practices that establish and support lifelong learning.